U0189618

Names of Undersea Features
in the Typical Areas of the Deep Sea and Ocean
Discovered and Proposed by China

深海典型海域
海底地理实体命名

Names of Undersea Features in the Typical Areas of the Deep Sea and Ocean Discovered and Proposed by China

石绥祥　樊　妙　邢　喆　主编

Chief Editors
Shi Suixiang, Fan Miao, Xing Zhe

中国海洋大学出版社
·青岛·

主 编

石绥祥 樊 妙 邢 喆

编委会

Chief Editors

Shi Suixiang, Fan Miao, Xing Zhe

Editorial Board

“海底地理实体”是洋床或海底的一部分，其具有可测量的地形起伏或由地形起伏划定其边界范围。海底地理实体命名，是对海底地理实体经过科学判别和认定后对其进行标识和命名的过程。随着海洋调查与开发能力的持续提升，海盆、海丘、海沟、海山等新的地理实体不断在海底被发现。对其命名不仅有助于精确定位海洋地理特征，而且可以建立一套共同的语言系统，使得海洋科学研究和海底资源开发等多项活动变得明确和便捷。海底地理实体名称承载着人类对海洋地理特征的探索和认知、对海洋文化和遗产的弘扬和传承、对海洋自然现象的敬畏和依赖。因此，开展海底地理实体命名不仅是科学活动的一部分，更是维护海洋生态平衡和促进人类与自然和谐共处的重要手段。

国际海底地名分委会（SCUFN）是履行由国际海道测量组织（IHO）和政府间海洋学委员会（IOC）共同赋予的国际海底地名命名职能的组织，每年举行一次工作会议，对提交的海底地名提案进行审议，审议通过的海底地名将写入世界海底地名录。在过去的 20 多年里，海底地名命名工作得到了世界各国海洋研究机构和科学家的普遍重视和积极参与。我国自 2009 年起积极参与 SCUFN 活动，迄今已连续参加 14 届 SCUFN 工作会议，提交并审议通过海底地名提案 261 个，公布我国管辖海域及有关海域海底地名近千个。

海底地名由通名和专名两个部分组成。其中，通名界定海底地理实体的属性；专名是地名的重要组成部分，由发现该海底地理实体的国家和个人提出，也是最能体现命名特色的内容，专名的选取需要从含义、适用性等多方面协调考虑，使其便于传播使用，并且能够弘扬本国文化。本书主要依托我国近五年来的多波束水深调查资料，对其进行数据精细化处理、质量控制和精度评估，在满足海底地名命名要求的基础上，依据《国际海底地名命名标准 IHO-B6》、GB 29432-2012《海底地名命名》，对西太平洋、东北太平洋、中北太平洋、东印度洋部分海域 78 个典型海底地理实体进行了判读、提取和命名，其中 18 个已被 SCUFN 收录至世界海底地名录中。

在充分吸纳、借鉴世界海底地名录及我国公布的海底地名的基础上，此次命名的海底地理实体专名，首先遵循指位性原则，考虑与其相联系的已有地理要素特征进行命名，其次以体现深远海地域、时代特色、中国文化特色为原则，采取团组化和系列化方式，初步形成深海特色海底地名命名规划体系。具体为，大型地理实体，如一、二级地理实体起到控制全局的作用，其专名尊重传统使用已久的名称，或以所在海域的相对位置或其邻近陆地地名命名，便于海底地理实体的搜索和识别；三、四级地理实体的命名以象形为主，充分体现深远海调查、中国特色及时代特色元素，传承中华文化，纪念古代杰出人物。

本书共包括五大部分，第一部分主要介绍本名录海底地理实体通名的术语和定义，第二至第五部分为上述四个海域的海底地理实体分布图及命名情况，每个海底地理实体由详细信息和二维、三维图共同描述其位置信息、命名依据、形态和特征。

为了使本书所载的海底地理实体的命名原则与国际上规定的命名原则保持一致，本书所载的海底地理实体的命名严格按照上述《国际海底地名命名标准 IHO B-6》、GB 29432-2012《海底地名命名》规定的海底地理实体命名的原则和标准进行编撰，并援引了其中所载的术语与定义。本书仅供科学研究和海洋调查以及作为一般参考资料使用。

本书已经过中国地名委员会海底地名分委会（CCUFN）的审议和批准。

在本书出版之际，感谢国家卫星海洋应用中心蒋兴伟院士、自然资源部第二海洋研究所李家彪院士、自然资源部第二海洋研究所郑玉龙研究员、自然资源部海洋预警监测司叶菁二级巡视员的大力支持，自然资源部南海调查中心高金耀研究员、广州海洋地质调查局朱本铎正高级工程师、民政部地名研究所阮文斌研究员等专家在通名界定和专名命名方面给予的技术指导和审核，以及中国常驻国际海底管理局代表处原副代表毛彬研究员对本书译文的悉心审校。同时，也衷心感谢海底地形地貌调查项目的负责人和参与者，没有他们艰苦的外业工作、严谨的内业处理，就没有得以支撑海底地名命名的基础资料。随着我们对海底结构和地形特征认识的不断深化，我们将继续开展其他海域的海底地名识别和命名。本书不足之处在所难免，恳请斧正。

《深海典型海域海底地理实体命名》编写组

2024 年 1 月

Preface

The "Undersea feature" is a part of the ocean floor or seabed that has measurable relief or is delimited by relief. The undersea feature naming is the process of assigning names to various geographical features and formations discovered on the ocean floor or seabed. With the continuous improvement of marine survey and development capabilities, new geographical features such as basins, seamounts, trenches, and hills are constantly discovered on the seabed or ocean floor. To name these undersea features not only helps to accurately locate marine geographical features but also establishes a common language system, making the conduction of the activities such as marine scientific research and seabed resource development with high efficiency. The undersea feature naming carries human exploration and cognition of marine geographical features, the promotion and inheritance of marine culture and heritage, and the reverence and dependence on marine natural phenomena. Therefore, undersea feature naming is part of scientific activities, as well as an important means to maintain marine ecological balance and promote harmonious coexistence between humans and nature.

The GEBCO Sub-committee on Undersea Feature Names (SCUFN) is an organization performing the function entrusted to it by IHO and IOC for undersea feature names. It holds an annual conference to select, considerate, review, and make advice to the undersea feature name proposals. The undersea feature names shall be reviewed by SCUFN before they are added to the Gazetteer of Undersea Feature Names. Over the past two decades, many marine research institutions and scientists all over the world have attached great importance to undersea feature names and actively participated in the undersea feature naming. Since 2009, China has actively participated in SCUFN activities and has participated in its 14 conferences, made 261 proposals for undersea feature names which have been selected by SCUFN. Moreover, China have published about a thousand undersea feature names for the sea areas within its national jurisdiction and also for the other sea areas. We have named a total of 78 undersea features newly discovered in this publication, and 18 have been added to the Gazetteer of Undersea Feature Names.

According to the undersea feature naming principles, a specific term followed by a generic term makes up a feature name. The generic term is to define the type of undersea features. The specific term is an important part of the feature name, which is proposed by the country and individual who discovered the undersea feature, and it is also the part that can best reflect the characteristics of the feature. The specific term needs to be carefully considered from various perspectives such as meaning, applicability, and cultural significance. It should be conducive to the spreading and accurate use of undersea feature names.

In order to ensure its quality, this publication of the undersea feature names has been prepared based on the latest multi-beam bathymetric data collected in the last five years. The data has been systematically processed and QA/QC checked. Based on the requirements for undersea feature naming, and in accordance with the standardization of Undersea Feature Names by the GEBCO Sub-Committee on Undersea Feature Names (B-6 document) (hereinafter referred to as the standardization) and GB 29432-2012 Nomenclature of Undersea Feature Names, China has named in this publication

78 undersea features newly discovered in the typical sea areas in the Western Pacific Ocean, Northeastern Pacific Ocean, Central North Pacific Ocean, and Eastern Indian Ocean, and 18 have been added to the Gazetteer of Undersea Feature Names.

With special reference to the guidance and methodology for naming the undersea features contained in the standardization, and the names of the undersea features published in China, the specific term is determined according to its location and already named features associated with it. At the same time, the specific term should reflect the sea area topography, characteristics of the times, and the characteristics of Chinese culture through adopting group names and serial names. The undersea features at the first and second levels are generally named after the sea areas where the features are located and the nearby continents and islands to reflect their location. The undersea features at the third and fourth levels are generally named after geometric shapes of the features, or the characteristics of region and times, or after the Chinese cultural elements to promote Chinese culture, or after famous ancient figures to honor the memory of them.

This publication is composed of five chapters. Chapter I mainly introduces the terms and definitions of generic names for undersea features selected in this publication. Chapters II to V describe in detail the distribution and names of the undersea features in the four aforementioned sea areas. Each undersea feature is described in detail along with two-dimensional and three-dimensional maps, describing its location information, naming basis, morphology, and characteristics.

In order to bring the undersea feature names in this publication into line with the international undersea feature naming principles and standards, this publication is compiled in restrict accordance with the guidance and process for undersea feature naming contained in the standardization, and the terms and definitions contained therein are cited. This publication is intended for scientific research and oceanographic surveys and is to be distributed as a general reference.

This proposal for the names of the undersea features has been reviewed and approved by China Subcommittee on Undersea Feature Names (CCUFN).

At the moment of the completion of this publication, we would like to express our sincere gratitude to Academician Jiang Xingwei of the National Satellite Ocean Application Service, Academician Li Jiabiao of the Second Institute of Oceanography of the Ministry of Natural Resources, Researcher Zheng Yulong of the same institute, and Deputy Director-General Ye Jing of the Ministry of Natural Resources Marina Warning and Monitoring Department for their invaluable support. Our thanks also extend to the experts specializing in undersea feature naming and geoscience, such as Researcher Gao Jinyao from the South China Sea Marine Survey Center of the Ministry of Natural Resources, Professor of engineering Zhu Benduo from the Guangzhou Marine Geological Survey of the Ministry of Natural Resources, and Researcher Ruan Wenbin from the Institute of Place Names Research of the Ministry of Civil Affairs for their insightful guidance. Additionally, we would like to thank Former Deputy Representative Mao Bin of the Permanent Mission of the People's Republic of China to the International Seabed Authority for his meticulous proofreading of the translation work. At the same time, we extend our sincere gratitude to all the project leaders and investigators of the undersea topographic survey for their painstaking field work and rigorous in-door data processing and analysis, which provides us with a profound scientific data and information basis for undersea features naming.

The Editorial Committee of *Names of Undersea Features in the Typical Areas of the Deep Sea and Ocean Discovered and Proposed by China*

January 2024

目录
Contents

1

术语与定义

Chapter I

Terms and Definitions

1.1 海盆

海底洼地，平而大体呈等维展布，范围大小不一。

Basin

A depression more or less equidimensional in plan and of variable extent.

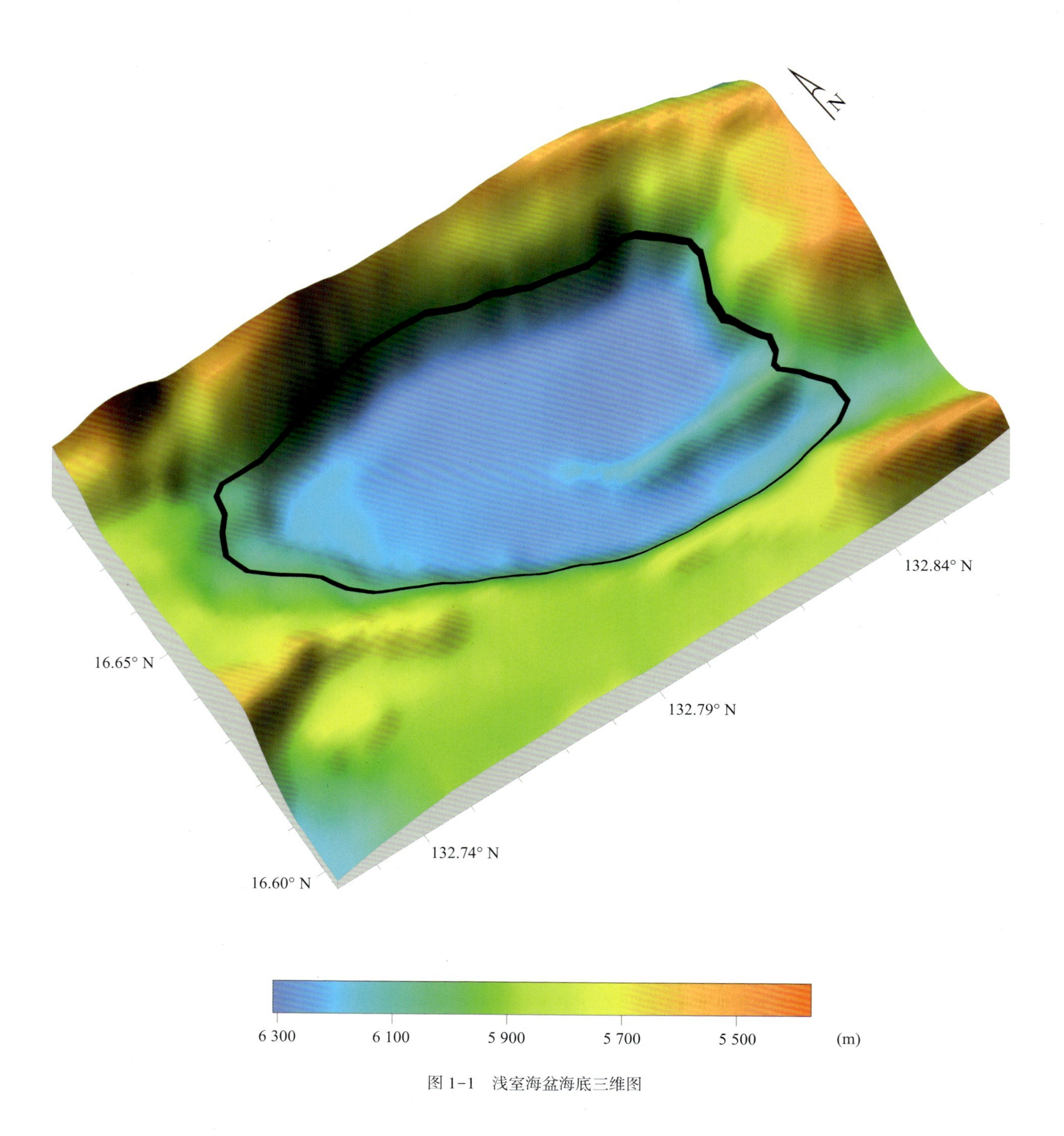

图 1-1 浅室海盆海底三维图

1.2 海渊

在较大实体区域内，如海槽、海盆或海沟中出现的局部深水区域。

Deep

A localized depression within the confines of a larger feature, such as a trough, basin or trench.

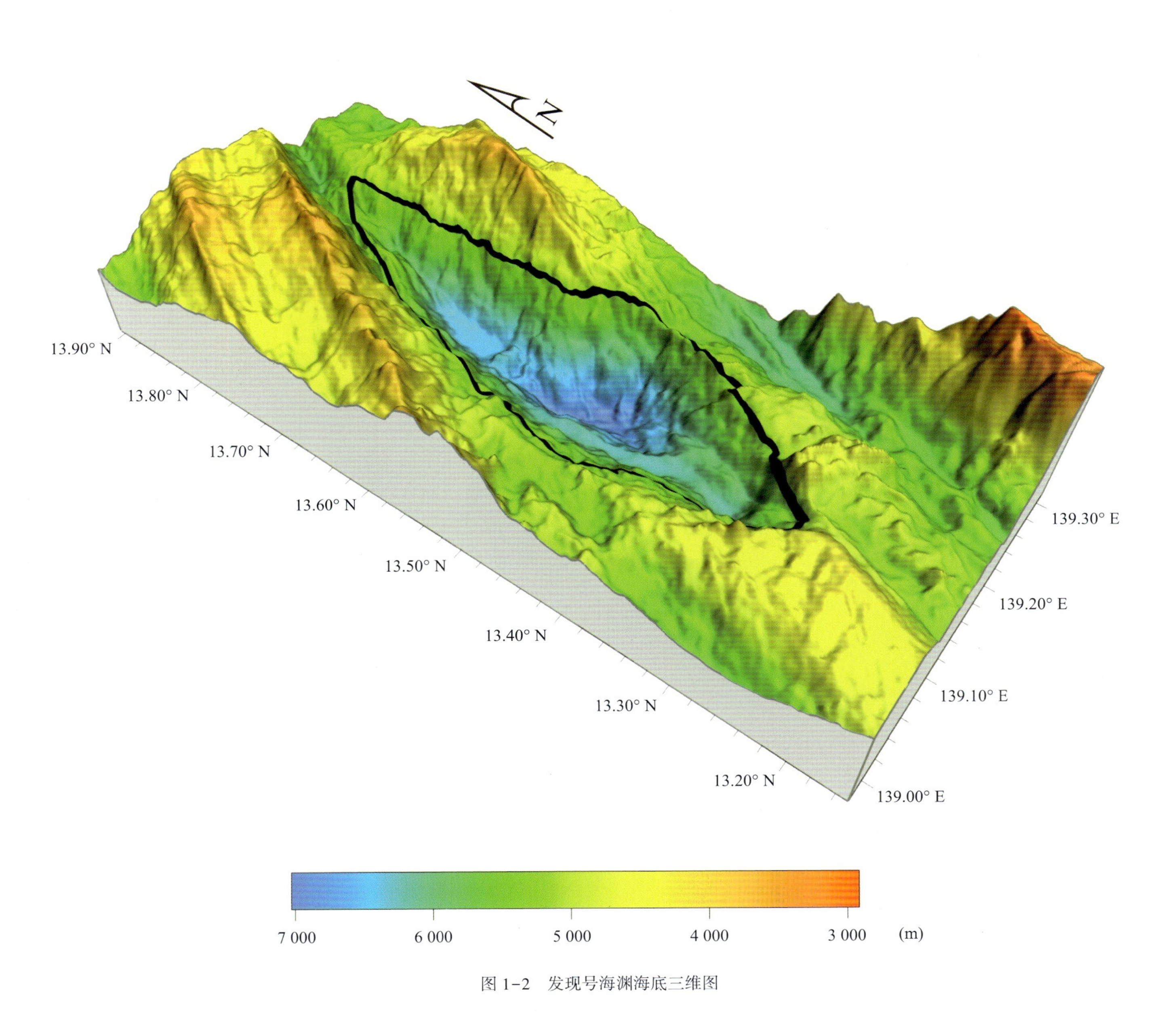

图 1-2 发现号海渊海底三维图

1.3 断裂带

因板块构造运动、伴随着海底扩张脊轴的位移而形成的狭长不规则地形区，区内常见有两翼陡峭和 / 或不对称的海脊、海槽或海底崖。

Fracture Zone

A long narrow zone of irregular topography formed by the movement of tectonic plates associated with an offset of a spreading ridge axis, characterized by steep-sided and/or asymmetrical ridges, troughs or escarpments.

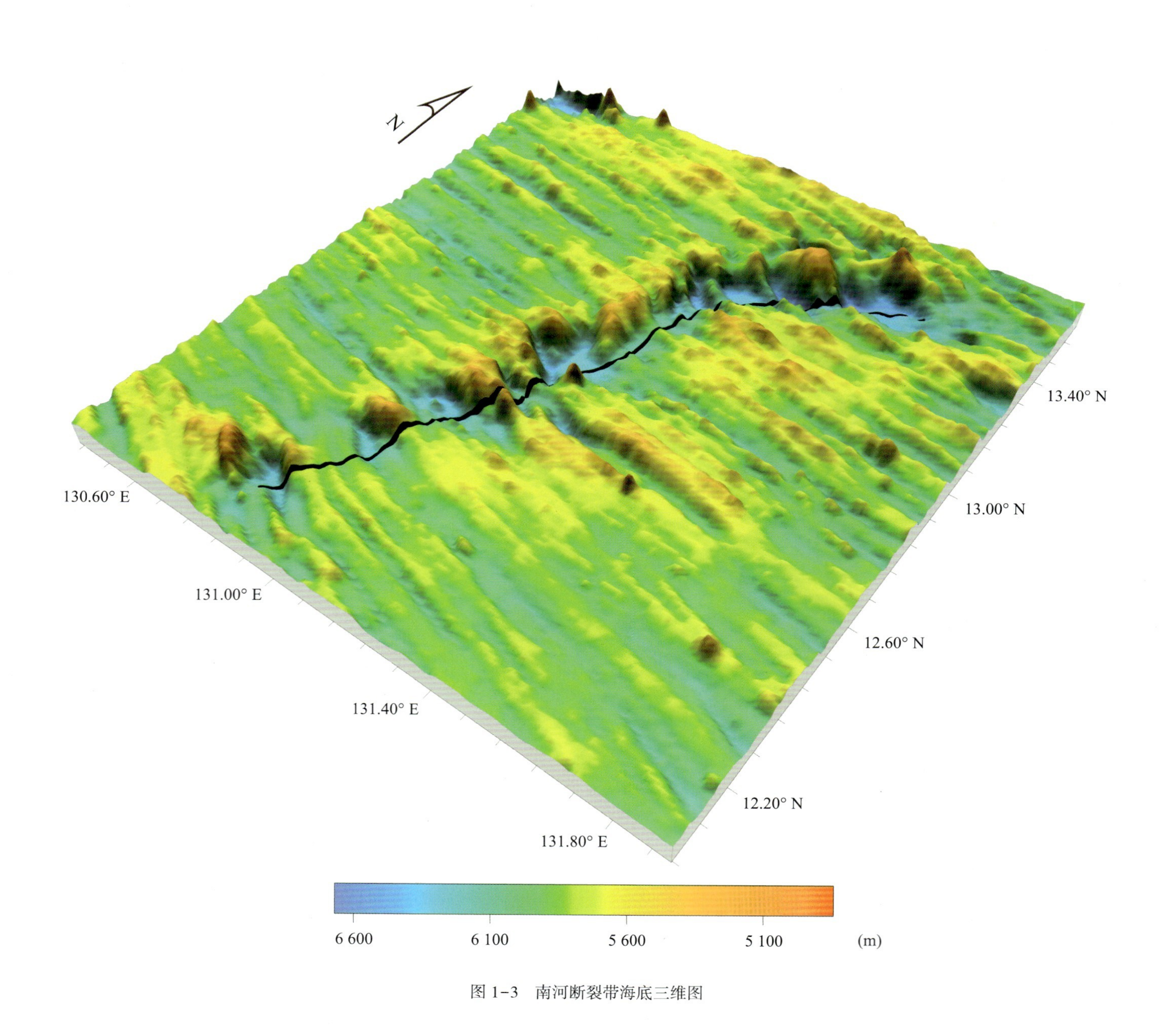

图 1-3　南河断裂带海底三维图

1.4 平顶山

顶部比较平坦的海山。

Guyot

A seamount with a comparatively smooth flat top.

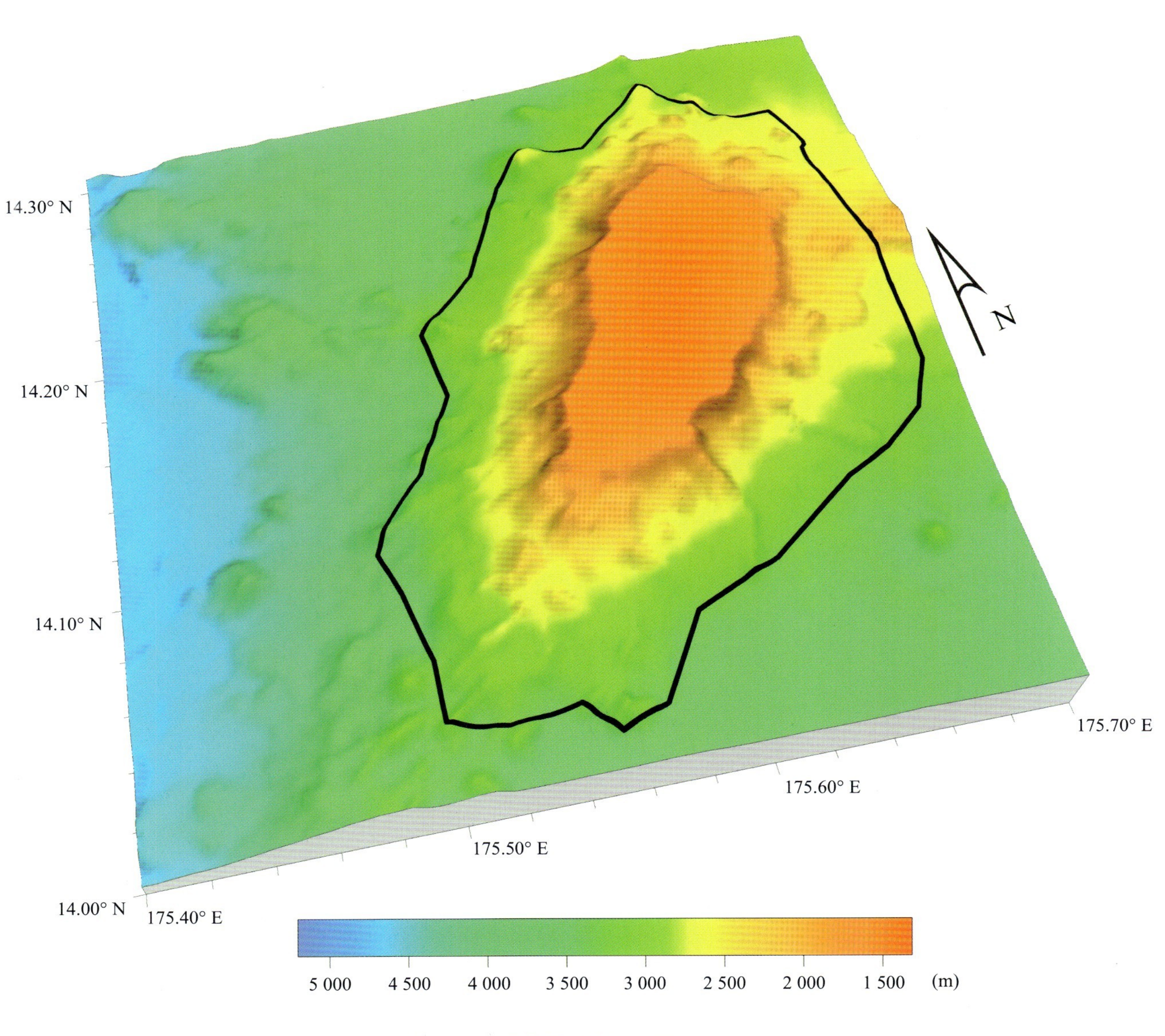

图 1-4 松骨平顶山海底三维图

1.5 平顶山群

由多个相对聚集的平顶山构成的大型地理实体。

Hill

A large undersea feature consisting of several relatively gathering guyots.

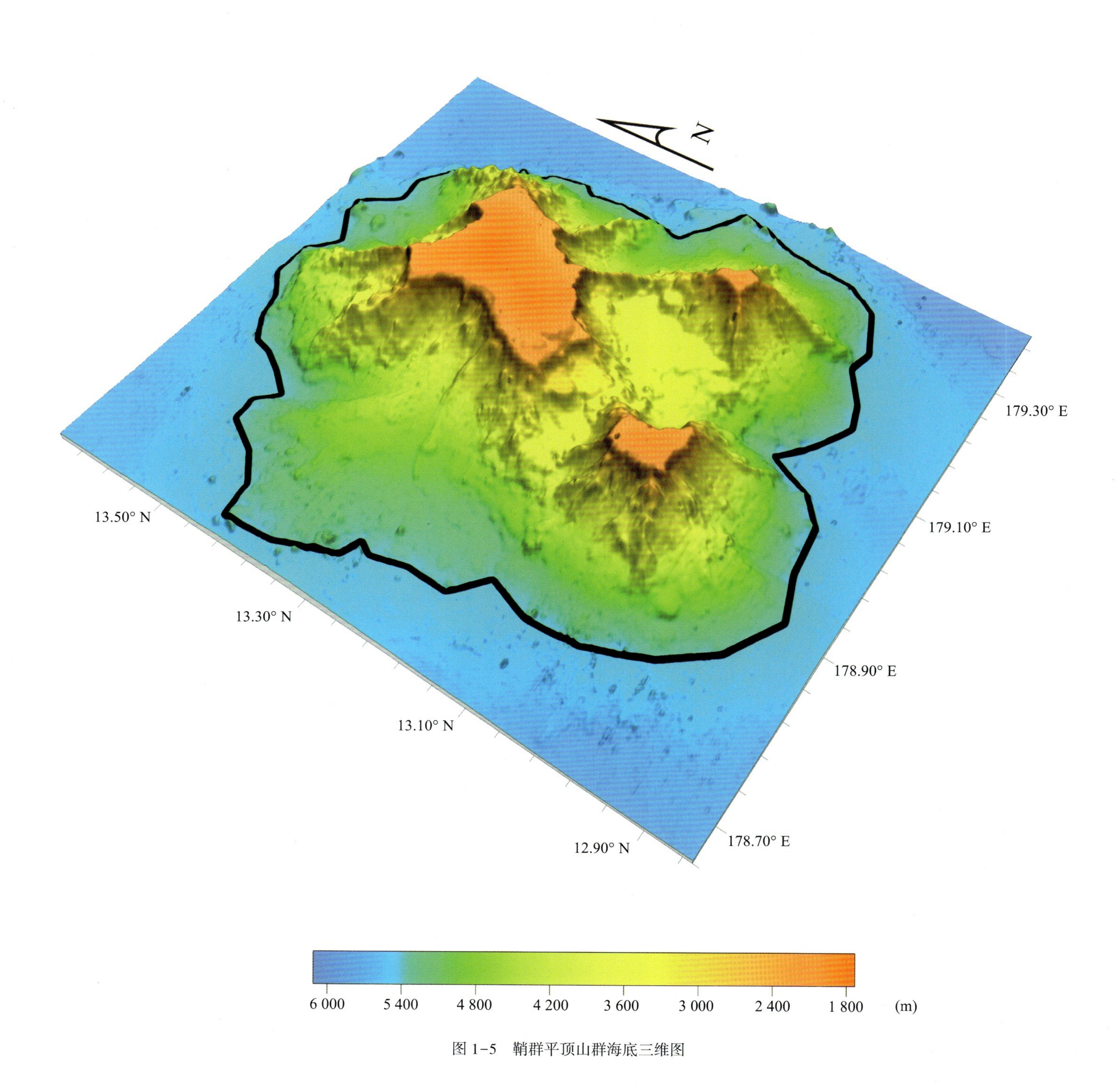

图 1-5　鞘群平顶山群海底三维图

1.6 海丘

清晰可辨的海底隆起区，形状一般不规则，从环绕其主体周围的最深等深线算起，顶部与周围地势起伏高差（相对高度）小于1 000 米。

Hill

A distinct elevation generally of irregular shape, less than 1,000 m above the surrounding relief as measured from the deepest isobath that surrounds most of the feature.

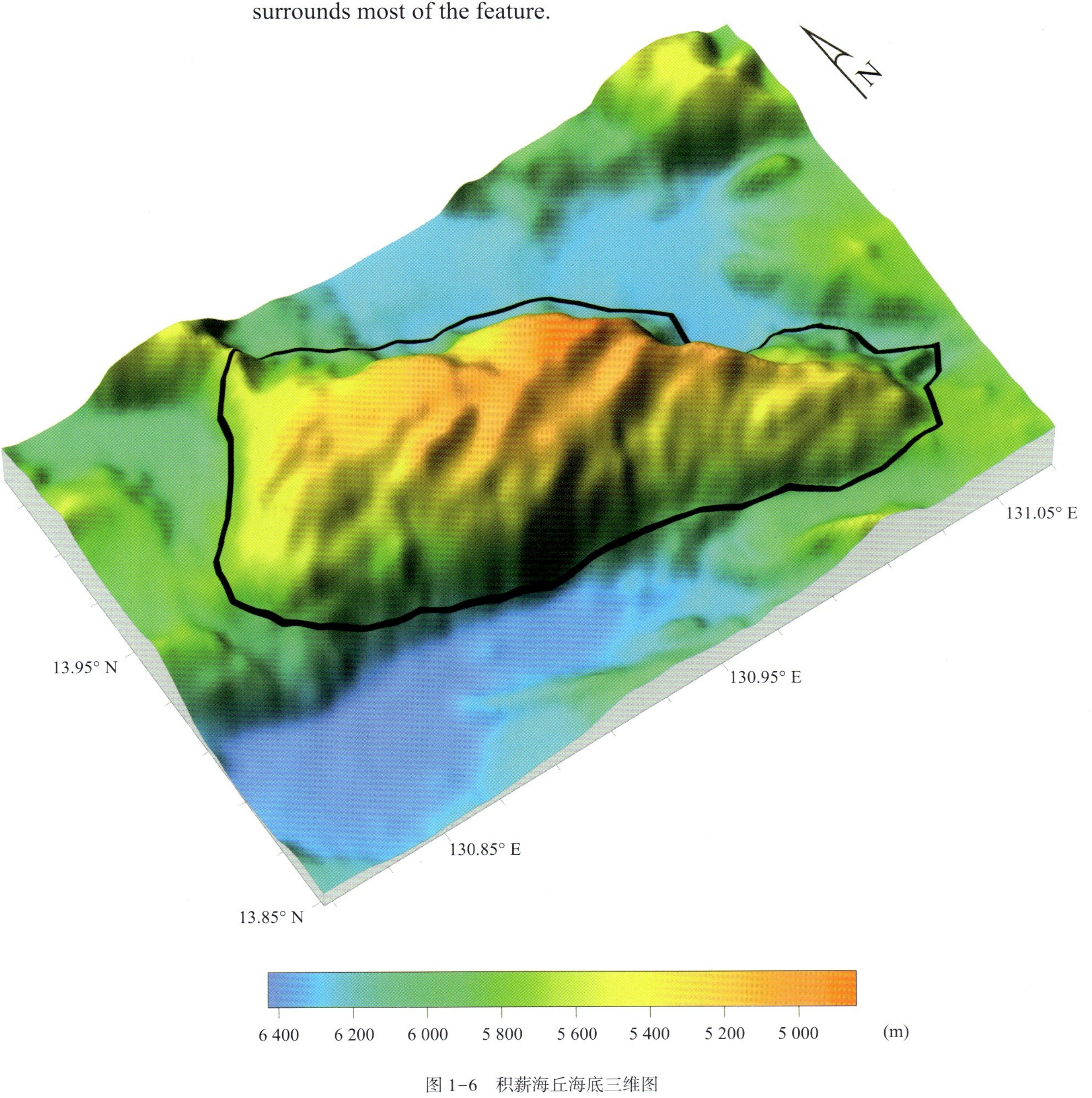

图 1-6 积薪海丘海底三维图

1.7 海丘群

由多个相对聚集的海丘构成的大型地理实体。

Hills

A large undersea feature consisting of several relatively gathering hills and/or knolls.

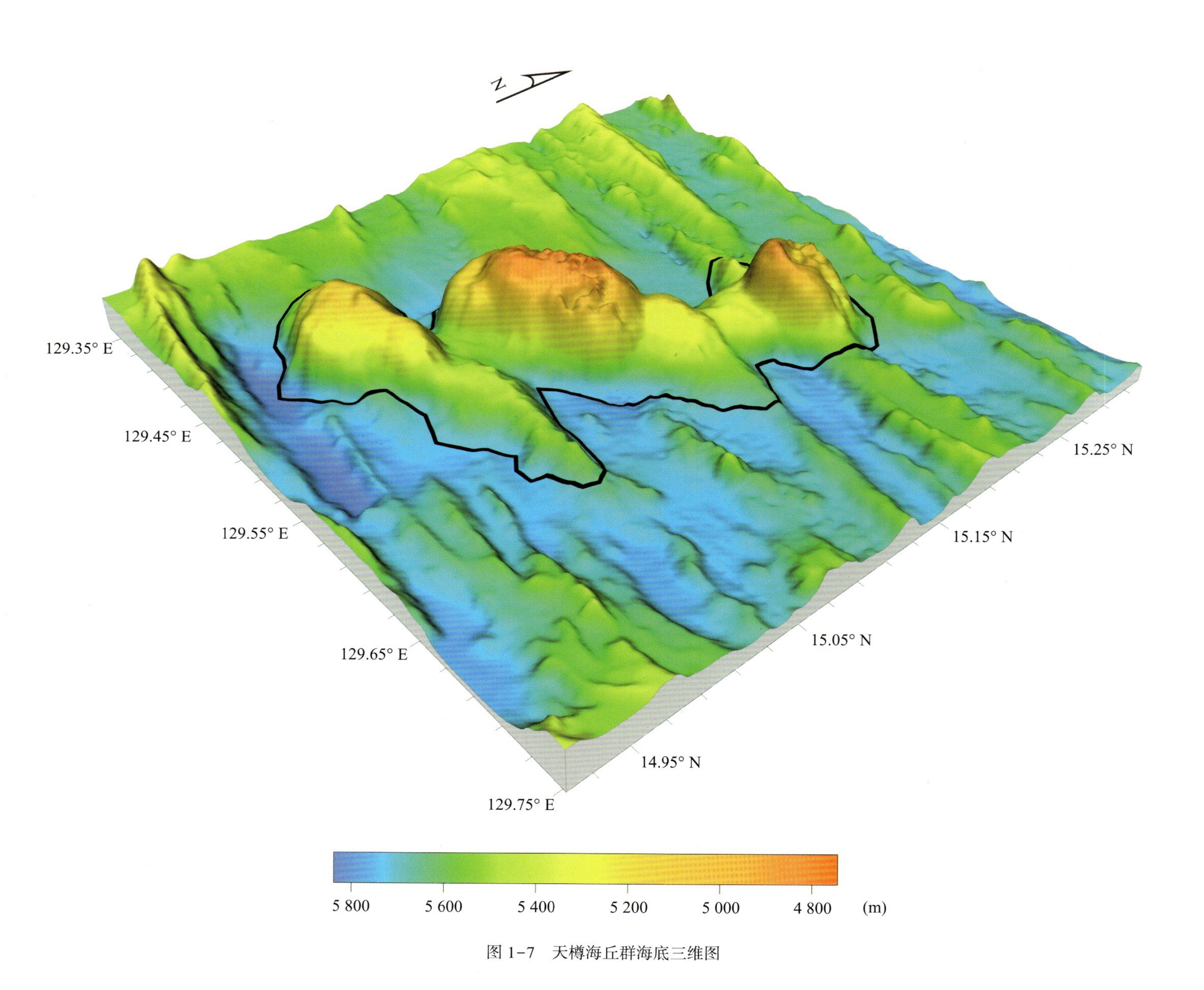

图 1-7 天樽海丘群海底三维图

1.8 圆丘

清晰可辨的海底隆起区，外轮廓呈圆形，从环绕其主体的最深等深线算起，顶部与周围地势起伏高差（相对高度）小于 1 000 米。

Knoll

A distinct elevation with a rounded profile less than 1,000 m above the surrounding relief as measured from the deepest isobath that surrounds most of the feature.

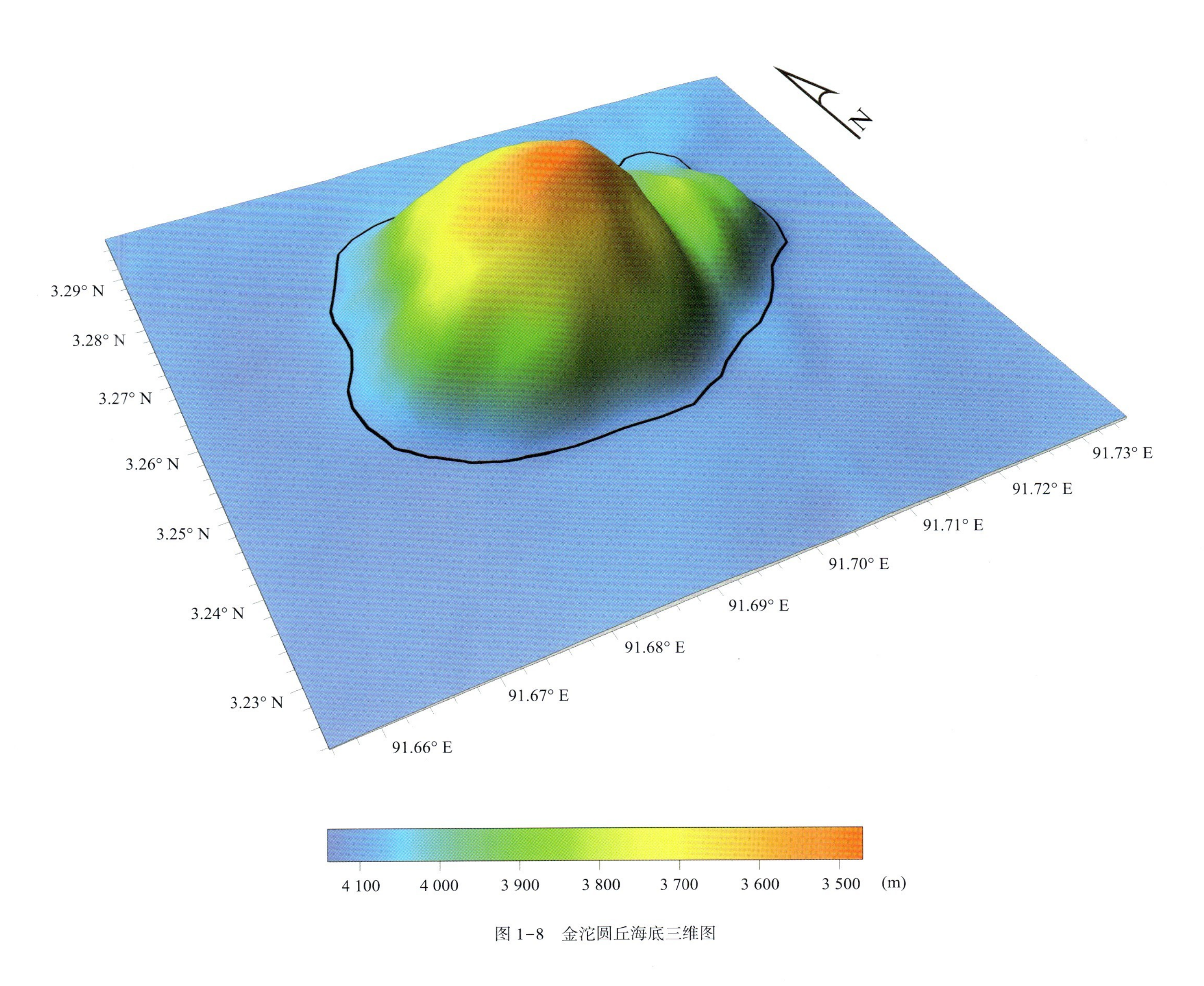

图 1-8 金沱圆丘海底三维图

1.9 海脊

大小和范围变化复杂的狭长隆起地带，通常具有陡峭的边坡。

Ridge

An elongated elevation of varying complexity and size, generally having steep sides.

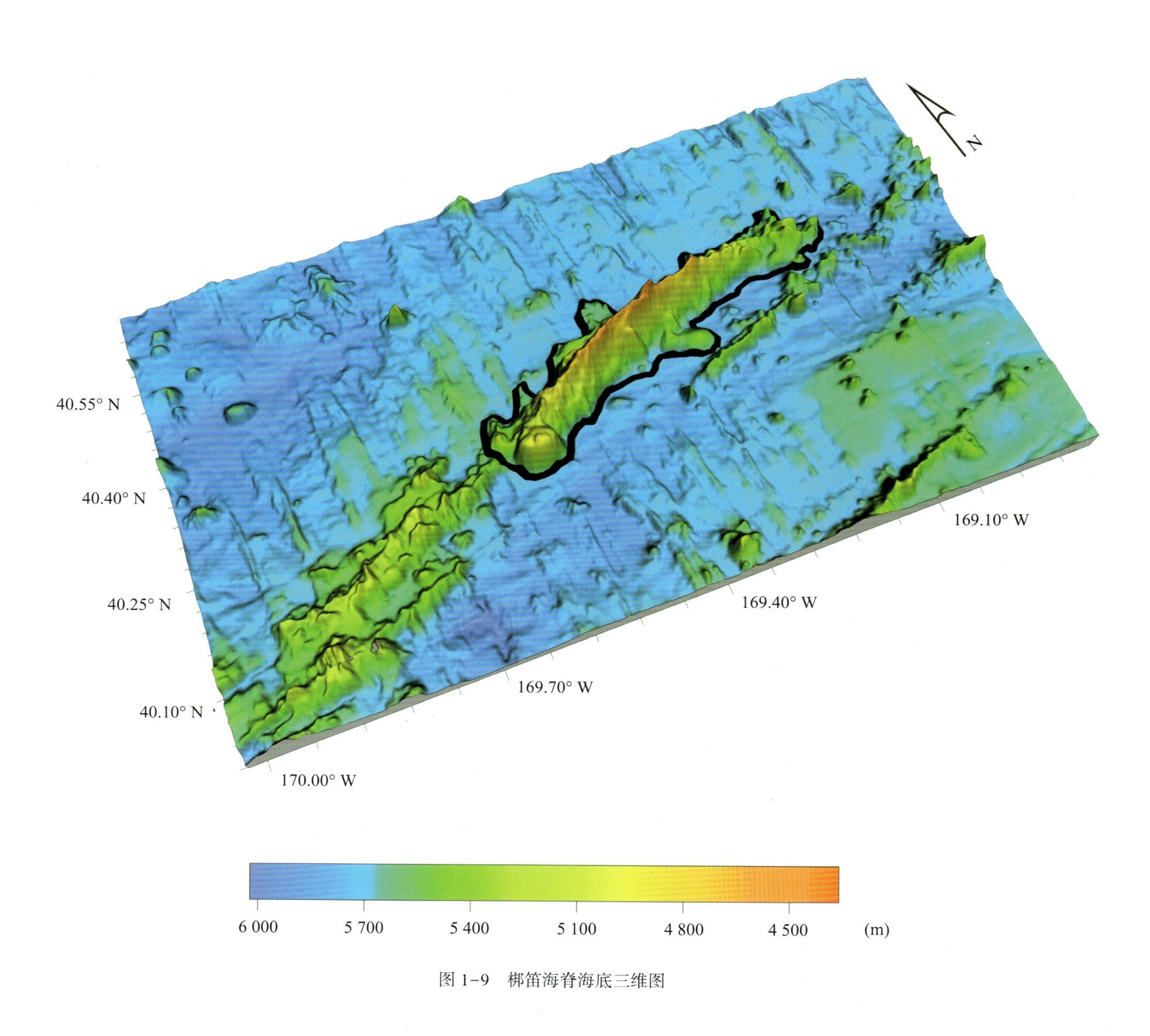

图 1-9 榔笛海脊海底三维图

1.10 海脊群

由多个相对聚集的海脊构成的大型地理实体。

Ridges

A large undersea feature consisting of multiple relatively gathering ridges.

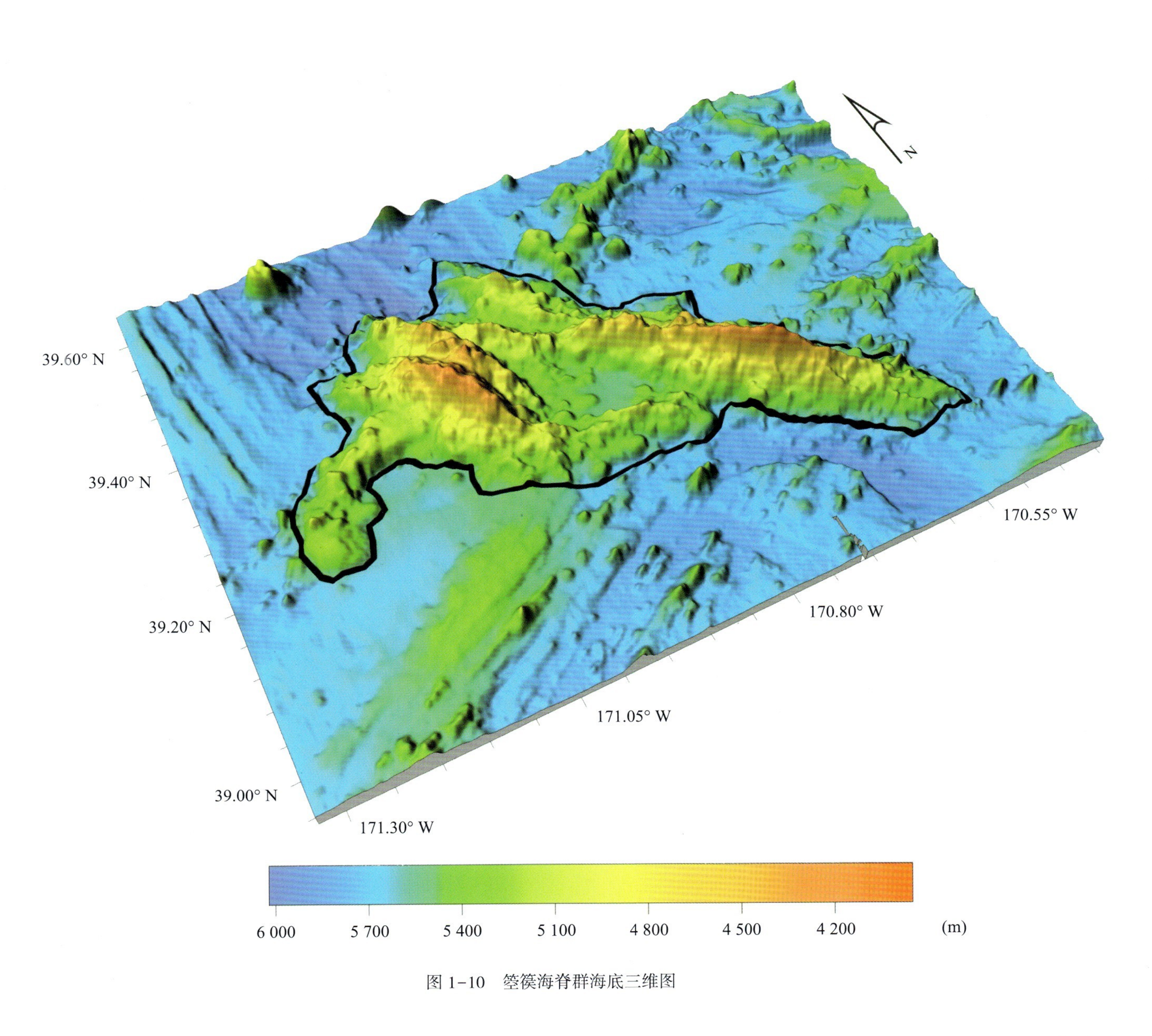

图 1-10 箜篌海脊群海底三维图

1.11 海底水道

细长而蜿蜒曲折的海底洼地，通常出现在平缓倾斜的平原或海扇中。

Sea Channel

An elongated, meandering depression, usually occurring on a gently sloping plain or fan.

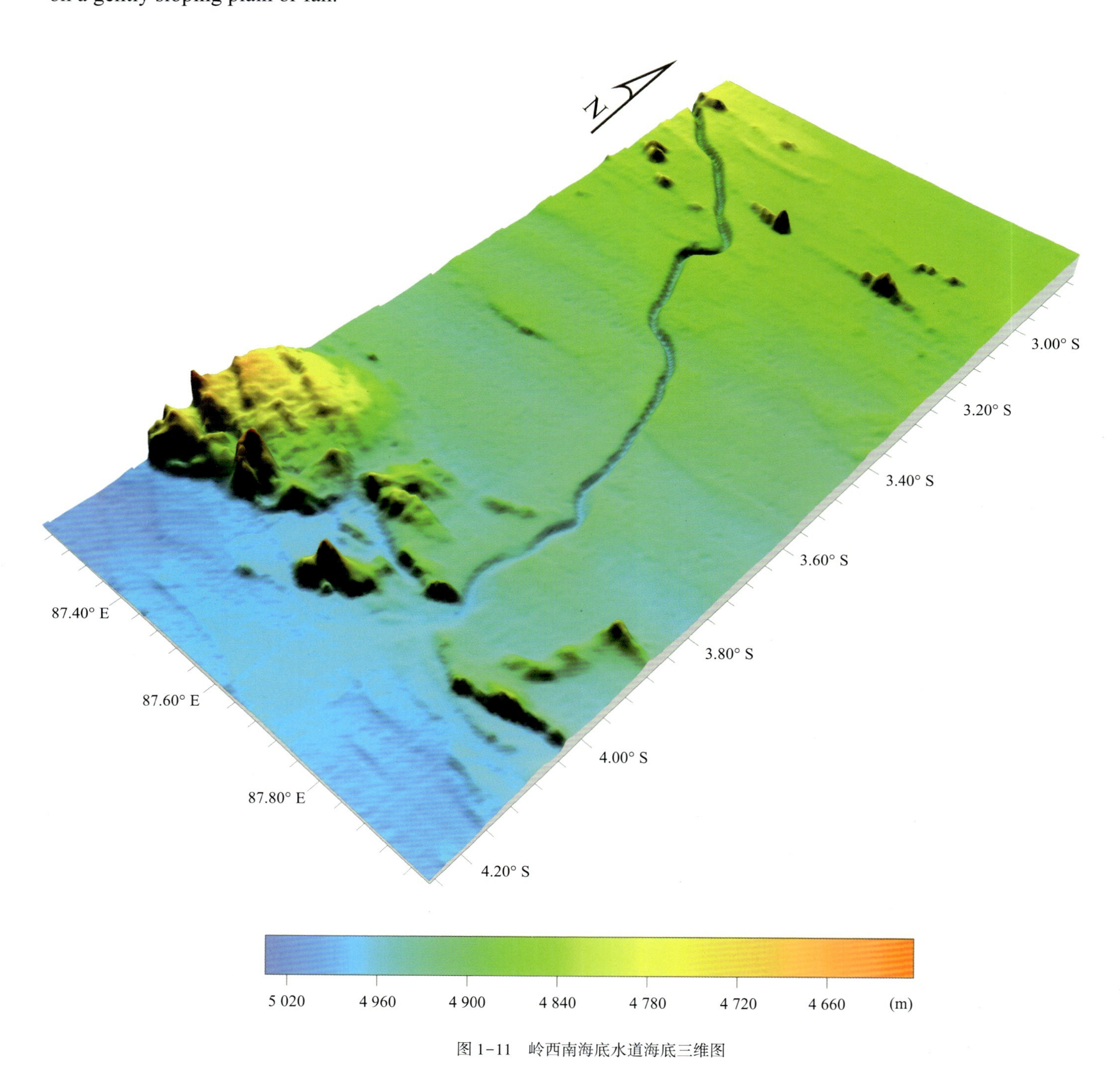

图 1-11 岭西南海底水道海底三维图

1.12 海山

清晰可辨的、大体呈等维展布的海底高地，从环绕其主体的最深等深线算起，顶部与周围地势起伏高差(相对高度)大于 1 000 米。

Seamount

A distinct generally equidimensional elevation greater than 1,000 m above the surrounding relief as measured from the deepest isobath that surrounds most of the feature.

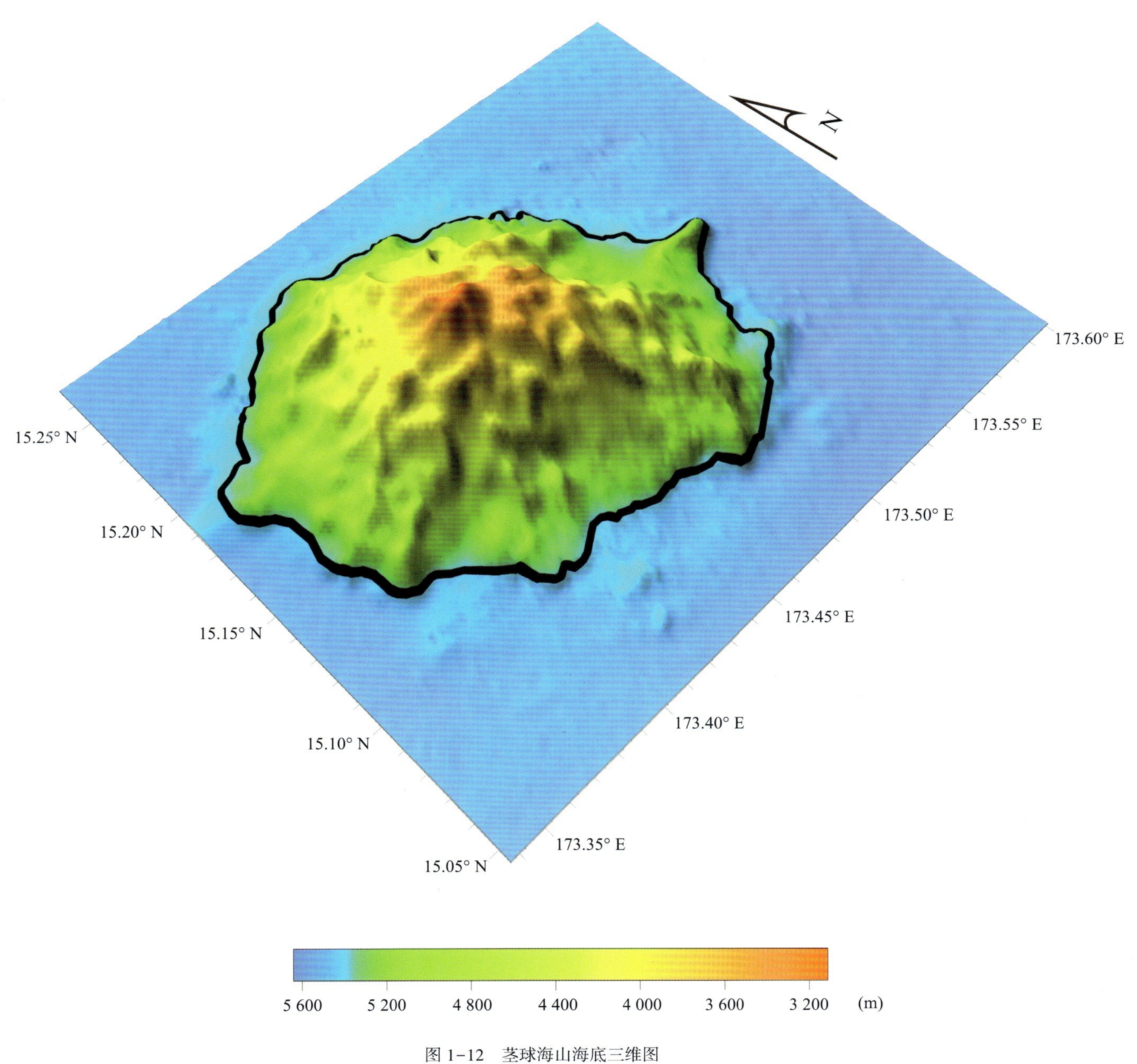

图 1-12　茎球海山海底三维图

1.13 海山群

由多个相对聚集的海山构成的大型地理实体。

Seamounts

A large undersea feature consisting of multiple relatively gathering seamounts.

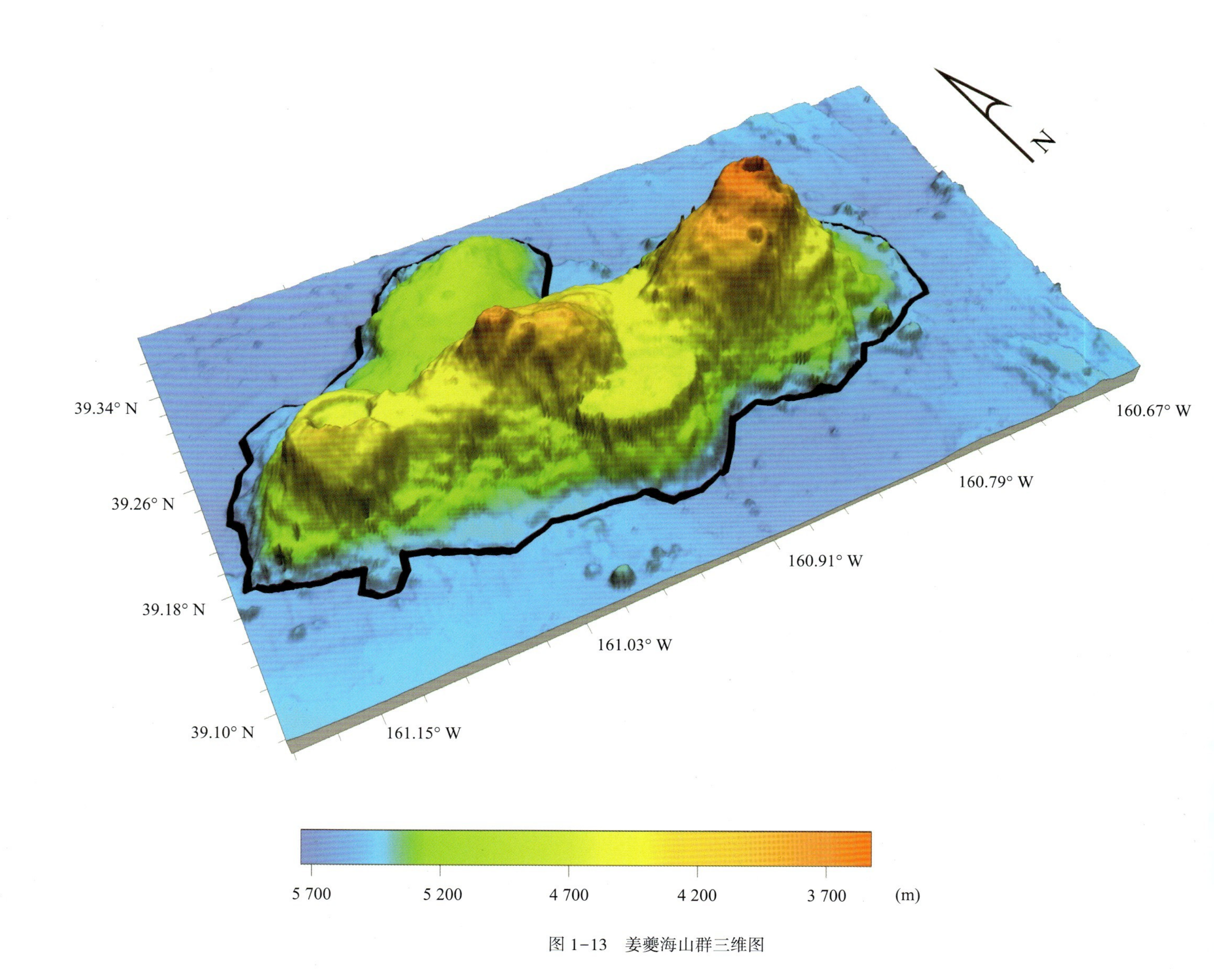

图 1-13 姜夔海山群三维图

2

西太平洋海底地理实体

Chapter II

Undersea Features in the Western Pacific Ocean

2.1 地形地貌概况

菲律宾海是西太平洋的重要组成部分，是西太平洋最大的边缘海。它位于太平洋西部菲律宾群岛的东面和北面，覆盖面积 5.4 × 10^6 km^2，北部位于 130°E~142°E 之间，南部处于 124°E~147°E 之间，南北纬度跨越 0°~35°N。周围被岛弧和海沟包围，海底地势的主要特征是南北长、北窄南宽。菲律宾海水深一般在 3 000~8 000 m 范围内，总体表现为“西部深，东部浅”的特点。菲律宾海板块受到西部亚欧板块、东部太平洋板块和东南侧加罗林板块三大板块的俯冲、碰撞，又经历了走滑断裂作用以及岛弧的形成和弧后扩张作用，海底地形复杂多样，海台、海山、海岭呈块状或链状散布在海盆之上，海脊与岛弧大致呈南北向伸展并贯穿全区，海岭和海盆并列。因此，菲律宾海是具有洋壳基底的大型边缘海。

Section 2.1 Overview of the topography

The undersea features named in this chapter are concentrated in the Philippine Sea in the Western Pacific Ocean. The Philippine Sea is an important part and the largest marginal sea of the Western Pacific Ocean. It is located to the east and north of the Philippine Islands in the western Pacific Ocean. It covers an area of 5.4 × 10^6 km^2, with the northern part encompassed between 130°E and 142°E and the southern part encompassed between 124°E and 147°E, with the latitudes spanning 35 degrees from 0° to 35°N. The Philippine Sea is generally a diamond shaped area that slightly stretches from south to north, surrounded by island arcs and trenches. The main features of the undersea terrain are that it is long from north to south, narrow in the north and wide in the south. The depths of the Philippine Sea generally range from 3,000 m to 8,000 m, with an overall characteristic of “deep in the west and shallow in the east”. The Philippine Sea Plate is subducted and collided by three major plates: the Eurasian Plate in the west, the Pacific Ocean Plate in the east, and the Caroline Plate in the southeast. It has also experienced strike-slip faulting, the formation of island arcs, and back-arc expansion. The seafloor topography is complex and diverse with plateaus, seamounts, and ridges scattering in blocks or chains on the ocean basin. The ridges and island arcs extend roughly in a north-south direction throughout the entire region, and the ridges and basins are juxtaposed.

Therefore, the Philippine Sea is a large marginal sea with an oceanic crust base.

2.2 菲律宾海地理实体命名

我国已对菲律宾海 30 个海底地理实体进行了命名，主要以月份的中文古称、二十四节气、象形特征等对海底地理实体进行命名。这些海底地理实体主要分布在九州－帕劳海脊以西海域，多沿中央海盆断裂带南北两侧分布。

考虑到与 SCUFN 命名体系的一致性，使用我国传统星官体系“三垣二十八星宿”[1]、西太平洋生物物种、海洋调查设备对该海域的海底地理实体进行命名。具体方案为，以九州－帕劳海脊为界，东、西、南、北四个方向以东方苍龙、西方白虎、南方朱雀、北方玄武中的星官进行命名，选词考虑了星官相对位置和海底地理实体相对位置的一一对应。此次命名的西太平洋海底地理实体主要分布在中央海盆断裂带以南和帕里西维拉（Parece Vela）海盆中部。

新命名的海底地理实体共 20 个，其中断裂带 5 个，为银河断裂带、北河断裂带、星河断裂带、南河断裂带、弧矢断裂带；海丘和海丘群 4 个，为积薪海丘、阙丘海丘、鱼尾海丘、天樽海丘群；海山和海山群 9 个，为半叶海山、摇光海山、开阳海山、玉衡海山、天权海山、天玑海山、天璇海山、天枢海山、梗河海山群；海盆和海渊各 1 个，为浅室海盆、发现号海渊。

1 三垣二十八星宿是中国古代天文学中的一个重要概念，用来描述北斗七星周围的星宿分布。其中，三垣是指紫微垣、太微垣和帝座垣，而二十八星宿则是围绕在三垣周围的星宿群，代表着不同的方向和季节。三垣二十八星宿在中国古代天文学中具有重要的象征意义，被用来测定时间、导航和预测天象等，它们也被广泛地应用于中国古代的历法和风水学中，对中国传统文化产生了深远的影响。

Section 2.2 Undersea features in the Philippine Sea

China has named 30 undersea features in the Philippine Sea, mainly after the ancient Chinese names of months, the twenty-four solar terms, pictographic characteristics, etc. The named undersea features are mainly distributed in the west of the Kyushu-Palau Ridge, mostly along the north and south sides of the central basin fault rift.

In order to make our naming principle consistent with that of the SCUFN naming system, the traditional Chinese star system “Three Enclosures and Twenty-Eight Constellations”[1], Western Pacific biological species, and marine survey equipment are used for naming the undersea features in this sea area. The specific plan is to take the Kyushu-Palau ridge as the boundary, and the four directions of east, west, south and north are named after the Eastern Canglong, the Western White Tiger, the Southern Vermilion Bird and the Northern Dance. The selection of words took into account the one-to-one correspondence between the relative positions of the constellations and the relative positions of the undersea feature. The newly discovered undersea features named in this publication are mainly distributed south of the Central Basin Fault Zone and in the central part of the Parece Vela Basin in the Western Pacific Ocean.

A total of twenty undersea features newly discovered in the Philippine basin named in this publication include five fracture zones: Yinhe Fracture Zone, Beihe Fracture Zone, Xinghe Fracture Zone, Nanhe Fracture Zone, Hushi Fracture Zone; four hills including one relatively gathering hills (hill group): Jixin Hill, Queqiu Hill, Yuwei Hill and Tianzun Hills (hill group); nine seamounts including one relatively gathering seamounts (seamount group): Banye Seamount, Yaoguang Seamount, Kaiyang Seamount, Yuheng Seamount, Tianquan Seamount, Tianji Seamount, Tianxuan Seamount, Tianshu Seamount and Genghe Seamounts (seamount group); one basin and one deep, namely, Qianshi Basin and Faxianhao Deep.

1 The Three Enclosures and Twenty-Eight Constellations is an important concept in ancient Chinese astronomy used to describe the distribution of stars around the Big Dipper. The Three Enclosures refer to the Purple Forbidden Enclosure, the Supreme Palace Enclosure, and the Heavenly Market Enclosure, while the Twenty-Eight Constellations are the star groups surrounding these enclosures, representing different directions and seasons. In ancient Chinese astronomy, the Three Enclosures and Twenty-Eight Constellations held significant symbolic importance, used for timekeeping, navigation, and celestial predictions. They were also widely applied in ancient Chinese calendrical systems and feng shui, exerting a profound influence on traditional Chinese culture.

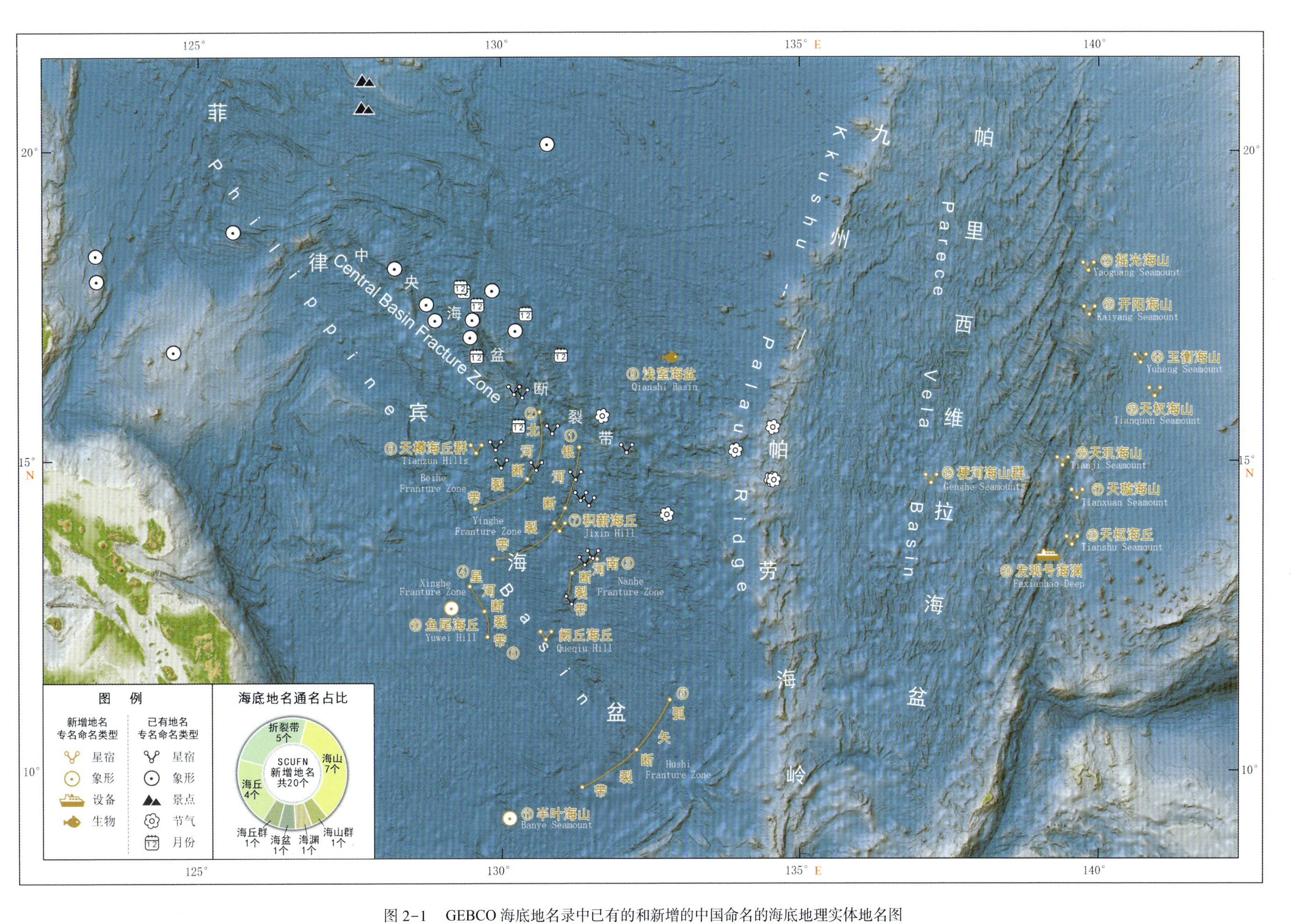

图 2-1 GEBCO 海底地名录中已有的和新增的中国命名的海底地理实体地名图

Fig.2-1 Map of the undersea feature names proposed by China and already included in GEBCO Gazetteer and these newly proposed by China

1	银河断裂带	Yinhe Fracture Zone	13°36.4′N 130°32.0′E
2	北河断裂带	Beihe Fracture Zone	15°03.2′N 130°41.7′E
3	南河断裂带	Nanhe Fracture Zone	13°17.8′N 131°22.9′E
4	星河断裂带	Xinghe Fracture Zone	12°58.5′N 129°28.8′E
5	弧矢断裂带	Hushi Fracture Zone	09°48.6′N 131°33.7′E
6	天樽海丘群	Tianzun Hills	15°12.2′N 129°35.3′E 15°06.6′N 129°32.2′E 15°00.7′N 129°27.9′E
7	积薪海丘	Jixin Hill	13°56.4′N 130°58.0′E
8	阙丘海丘	Queqiu Hill	12°11.6′N 130°44.6′E
9	浅室海盆	Qianshi Basin	16°39.5′N 132°49.4′E
10	鱼尾海丘	Yuwei Hill	12°37.7′N 129°09.9′E
11	半叶海山	Banye Seamount	09°14.3′N 130°07.9′E
12	摇光海山	Yaoguang Seamount	18°08.3′N 139°50.0′E
13	开阳海山	Kaiyang Seamount	17°25.6′N 139°50.9′E
14	玉衡海山	Yuheng Seamount	16°39.8′N 140°42.5′E
15	天权海山	Tianquan Seamount	16°06.8′N 140°56.7′E
16	天玑海山	Tianji Seamount	14°59.9′N 139°23.9′E
17	天璇海山	Tianxuan Seamount	14°28.1′N 139°38.5′E
18	天枢海山	Tianshu Seamount	13°43.3′N 139°33.5′E
19	梗河海山群	Genghe Seamounts	14°42.7′N 137°11.8′E 14°42.6′N 137°05.4′E 14°41.5′N 137°07.3′E 14°43.8′N 137°17.4′E
20	发现号海渊	Faxianhao Deep	13°29.2′N 139°09.7′E

2.2.1 银河断裂带 Yinhe Fracture Zone

中文名称 Chinese Name	Yínhé Duànlièdài 银河断裂带		
英文名称 English Name	Yinhe Fracture Zone		
地理区域 / Location	西太平洋菲律宾海盆 The West Philippine Basin		
特征点坐标 Coordinate	13°36.4′N 130°32.0′E	长度 / Length	300 km
		宽度 / Width	16 km
水深 / Depth	4 261~6 524 m	高差 / Total Relief	2 263 m
发现情况 Discovery Facts	此断裂带于 2014 年“向阳红 10”船，2017 年“向阳红 19”船、“向阳红 06”船在执行航次调查时发现。 The fracture zone was firstly discovered in 2014 during the survey cruises carried out onboard the Chinese R/V *Xiang Yang Hong 10* and then in 2017 during the survey cruises carried out onboard R/V *Xiang Yang Hong 19* and R/V *Xiang Yang Hong 06* respectively.		
地形特征 Feature Description	银河断裂带位于菲律宾海盆中部，中央海盆断裂带（Central Basin Fracture Rift）以南，呈 NE-SW 向弧状延伸，南北长 300 km，东西宽 16 km。该断裂带底部表现为隆起、间断、挤压合并而交替出现的海脊、海底凹陷、海底深槽地貌，水深范围 4 261~6 524 m，最大高差 2 263 m，最深点位于断裂带中南部。 Located in the middle of the West Philippine Basin, and to the south of Central Basin Fracture Rift, Yinhe Facture Zone extends in the NE-SW direction as an arc, and it is 300 km long from south to north and 16 km wide from west to east. At the bottom of the fracture zone, the rising, disconnected and squeezed places are shown as ridges, depressions and deep grooves alternately. The depths range from 4,261 m to 6,524 m, with a maximum total relief of 2,263 m. The deepest point is located in the central south of the fracture zone.		
专名释义 Reason for Choice of Name	银河系在夜空中看起来像一条银白色的亮带，古人想象为天上的一条河，该断裂带呈弧状弯曲，犹如璀璨星辉蜿蜒而成的一条河道，故取银河为名。 The Yinhe (The Milky Way) looks like a silver-white bright band in the night sky. Our ancestors imagined that it was a river in the sky. The fracture zone bends like an arc, just like a meandering river formed by bright stars, hence the feature is named Yinhe Fracture Zone.		

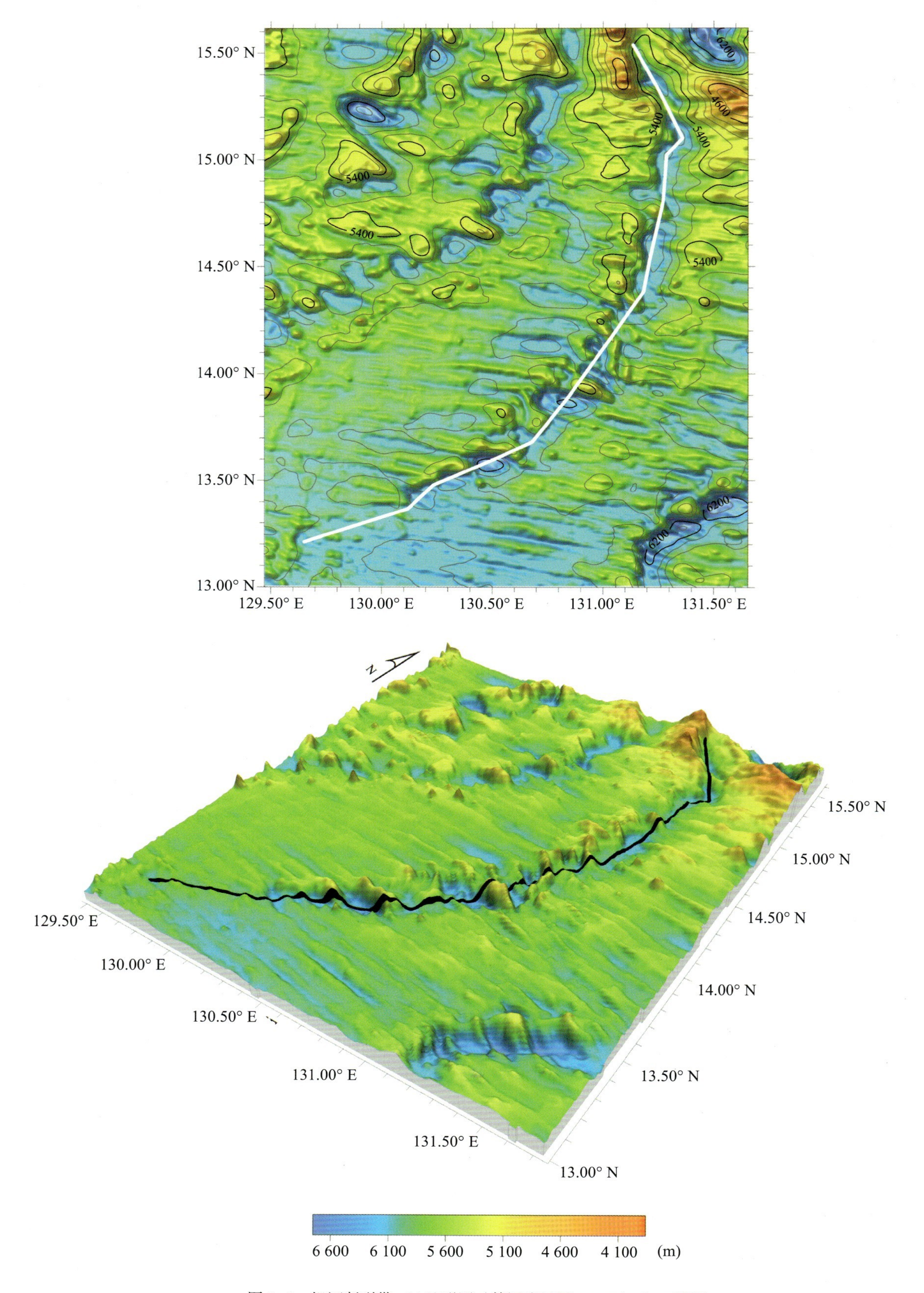

图 2-2　银河断裂带：(a) 地形图（等深线间隔 200 m）；(b) 三维图

Fig.2-2　Yinhe Fracture Zone: (a) Bathymetric map (Contours are in 200 m); (b) 3-D topographic map

2.2.2 北河断裂带 Beihe Fracture Zone

中文名称 Chinese Name	Běihé Duànlièdài 北河断裂带		
英文名称 English Name	Beihe Fracture Zone		
地理区域 / Location	西太平洋菲律宾海盆 The West Philippine Basin		
特征点坐标 Coordinate	15°03.2′N 130°41.7′E	**长度 / Length**	309 km
		宽度 / Width	17 km
水深 / Depth	4 534~6 409 m	**高差 / Total Relief**	1 875 m
发现情况 Discovery Facts	此断裂带于 2006 年“向阳红 14”船、2017 年“向阳红 19”船在执行航次调查时发现。 The fracture zone was firstly discovered in 2006 and then in 2017 during the survey cruises carried out onboard the Chinese R/V *Xiang Yang Hong 14 and* R/V *Xiang Yang Hong 19,*respectively.		
地形特征 Feature Description	北河断裂带位于菲律宾海盆中部，中央海盆断裂带以南，整体呈 NE-SW 向弧状延伸，南北长 309 km，东西宽 17km。该断裂带底部表现为隆起、间断、挤压合并而交替出现的海脊、海底凹陷、海底深槽地貌，水深范围 4 534~6 409 m，最大高差 1 875m，最深点位于断裂带中北部。 Located in the middle of the West Philippine Basin and to the south of Central Basin Fracture Rift, Beihe Fracture Zone extends in the NE-SW direction like an arc as a whole, and is 309 km long from south to north and about 17 km wide from west to east. At the bottom of the fracture zone, the rising, disconnected and squeezed places are shown as ridges, depressions and deep grooves alternately. The depths range from 4,534 m to 6,409 m, with a maximum total relief of 1,875 m. The deepest point is located in the central north of the fracture zone.		
专名释义 Reason for Choice of Name	北河，是我国传统星官体系“三垣二十八宿”南方朱雀第一宿井宿中的星官，位于现代星座划分的双子座。在古代星宿中，北河原是“河戎”，即河边的守卫，银河从南河和北河之间穿过，北河是银河北边的守卫。该断裂带在地形方位上位于银河断裂带的北面，故以北河命名。 The Beihe, as an asterism (or constellation) in the first constellation Jing Xiu (Well Constellation) of Nanfang Zhuque in the Three Enclosures and Twenty-Eight Constellations within the traditional Chinese constellation system, is part of Gemini according to the modern constellation system. In the ancient Chinese constellation system, the Beihe originally referred to “He Rong”, which meant a guard on the river bank. The Yinhe (The Milky Way) runs between Nanhe and Beihe, which is the guard in the north of the Yinhe. As the fracture zone is to the north of Yinhe Fracture Zone, hence the feature is named Beihe Fracture Zone.		

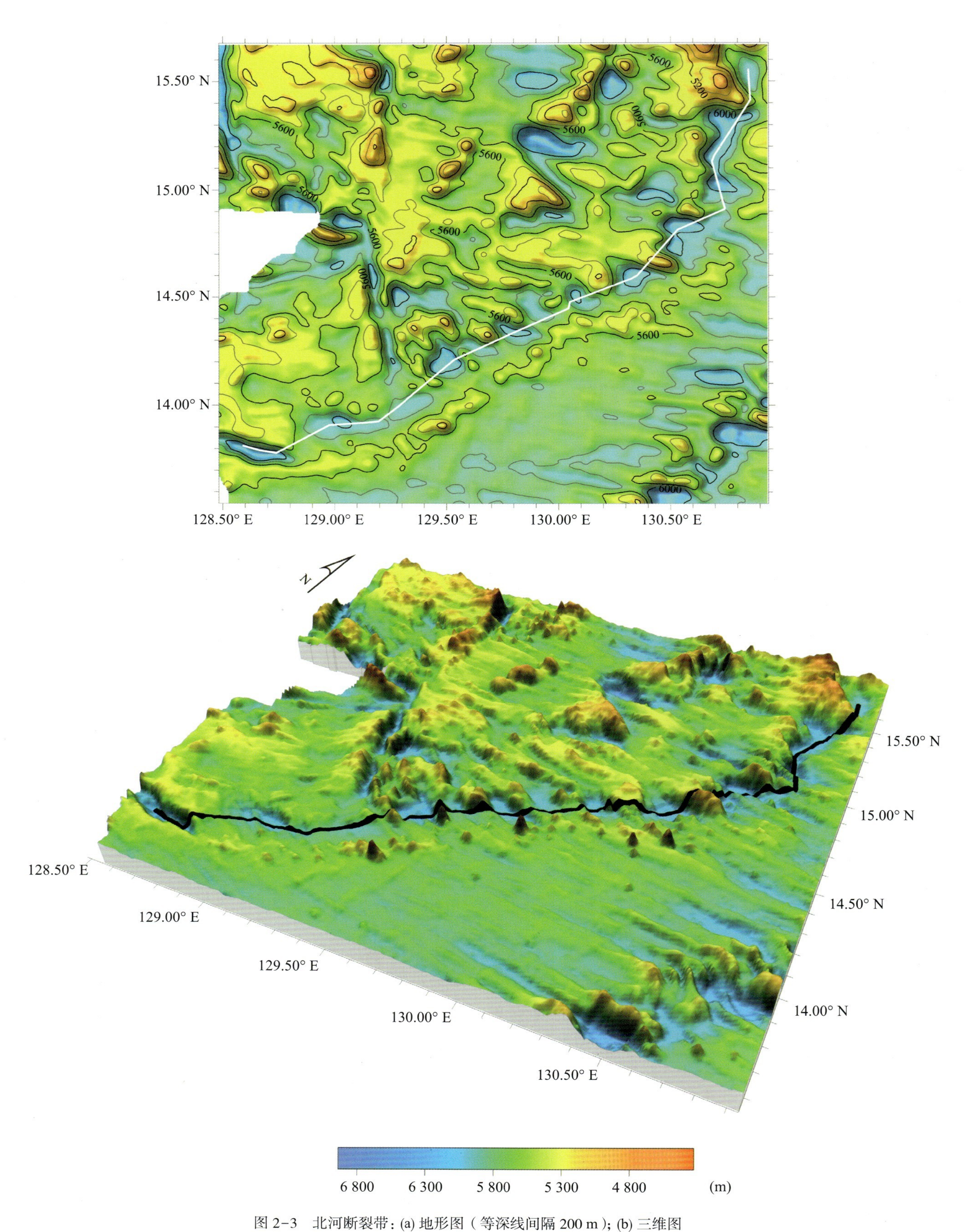

图 2-3 北河断裂带：(a) 地形图（等深线间隔 200 m）；(b) 三维图

Fig.2-3 Beihe Fracture Zone: (a) Bathymetric map (Contours are in 200 m); (b) 3-D topographic map

2.2.3 南河断裂带 Nanhe Fracture Zone

<table>
<tr><td>中文名称
Chinese Name</td><td colspan="3">Nánhé duànlièdài
南河断裂带</td></tr>
<tr><td>英文名称
English Name</td><td colspan="3">Nanhe Fracture Zone</td></tr>
<tr><td>地理区域 / Location</td><td colspan="3">西太平洋菲律宾海盆 The West Philippine Basin</td></tr>
<tr><td rowspan="2">特征点坐标
Coordinate</td><td rowspan="2">13°17.8′N 131°22.9′E</td><td>长度 / Length</td><td>163 km</td></tr>
<tr><td>宽度 / Width</td><td>20 km</td></tr>
<tr><td>水深 / Depth</td><td>5 050~6 663 m</td><td>高差 / Total Relief</td><td>1 613 m</td></tr>
<tr><td>发现情况
Discovery Facts</td><td colspan="3">此断裂带于 2017 年“向阳红 06”船在执行航次调查时发现。
The fracture zone was discovered in 2017 during the survey cruise carried out onboard the Chinese R/V Xiang Yang Hong 06.</td></tr>
<tr><td>地形特征
Feature Description</td><td colspan="3">南河断裂带位于菲律宾海盆中部，中央海盆断裂带以南，呈 NE-SW 向弧状延伸，南北长 163 km，东西宽 20 km。该断裂带底部表现为隆起、间断、挤压合并而交替出现的海脊、海底凹陷、海底深槽地貌，水深范围 5 050~6 663 m，最大高差 1 613 m。
Located in the middle of the West Philippine Basin and to the south of Central Basin Fracture Rift, Nanhe Fracture Zone extends in the NE-SW direction like an arc, and is 163 km long from south to north and about 20 km wide from west to east. At the bottom of the fracture zone, the rising, disconnected and squeezed places are shown as ridges, depressions and deep grooves alternately. The depth ranges from 5,050 m to 6,663 m, with a maximum total relief of 1,613 m.</td></tr>
<tr><td>专名释义
Reason for Choice of Name</td><td colspan="3">南河，是我国传统星官体系“三垣二十八宿”南方朱雀第一宿井宿中的星官，位于现代星座划分的小犬座。在古代星宿中，南河原是“河戍”，即河边的守卫，银河从南河和北河之间穿过，南河是银河南边的守卫。该断裂带在地形方位上位于银河断裂带的南面，和西北部北河断裂带一起组成重要的渡口。
The Nanhe, as an asterism in the first constellation Jing Xiu (Well Constellation) of Nanfang Zhuque in the Three Enclosures and Twenty-Eight Constellations within the traditional Chinese constellation system, is part of Canis Minor according to the modern constellation system. In the ancient Chinese constellation system, Nanhe originally referred to “He Rong”, which meant a guard on the river bank. The Yinhe runs between Nanhe and Beihe, and Nanhe refers to the guard in the south of the Yinhe. This fracture zone is to the south of Yinhe Fracture Zone, and serves as an important port together with Beihe Fracture Zone to the northwest.</td></tr>
</table>

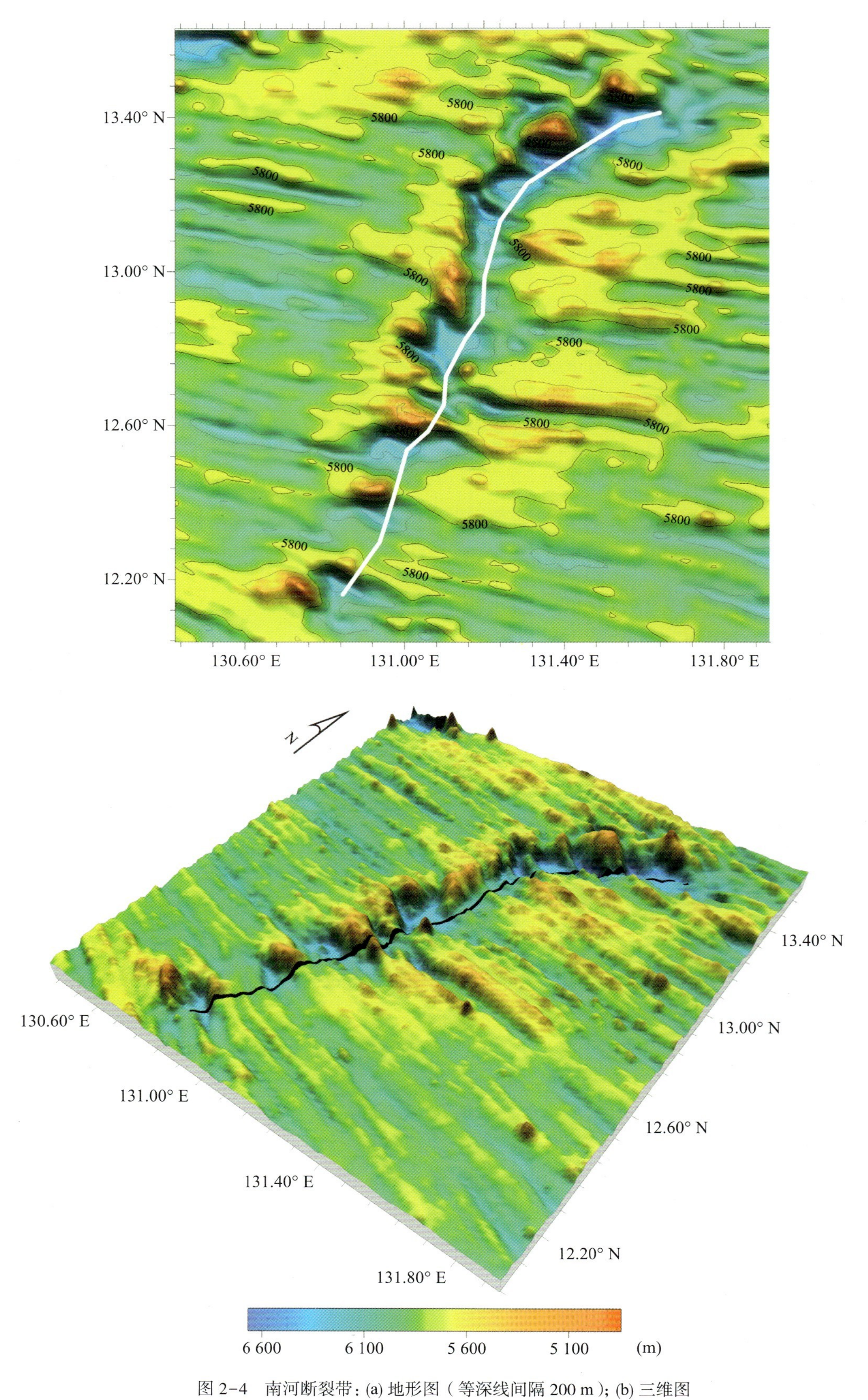

图 2-4　南河断裂带：(a) 地形图（等深线间隔 200 m）；(b) 三维图

Fig.2-4　Nanhe Fracture Zone: (a) Bathymetric map (Contours are in 200 m); (b) 3-D topographic map

2.2.4 星河断裂带 Xinghe Fracture Zone

中文名称 Chinese Name	Xīnghé Duànlièdài 星河断裂带		
英文名称 English Name	Xinghe Fracture Zone		
地理区域 / Location	西太平洋菲律宾海盆 The West Philippine Basin		
特征点坐标 Coordinate	12°58.5′N 129°28.8′E	长度 / Length	155 km
		宽度 / Width	11 km
水深 / Depth	5 225~6 206 m	高差 / Total Relief	981 m
发现情况 Discovery Facts	此断裂带于 2017 年“向阳红 06”船在执行航次调查时发现。 The fracture zone was discovered in 2017 during the survey cruise carried out onboard the Chinese R/V *Xiang Yang Hong 06*.		
地形特征 Feature Description	星河断裂带位于菲律宾海盆中部，中央海盆断裂带以南 385 km，呈 NW-SE 向弧状延伸，南北长 155 km，东西宽 11 km。该断裂带底部表现为隆起、间断、挤压合并而交替出现的海脊、海底凹陷、海底深槽地貌，水深范围 5 225~6 206 m，最大高差 981 m，最深点位于断裂带中北部。 Located in the middle of the West Philippine Basin, the fracture zone is about 385 km south of Central Basin Fracture Rift. Xinghe Fracture Zone extends in the NE-SE direction like an arc, and is 155 km long from south to north and about 11 km wide from west to east. At the bottom of the fracture zone, the rising, disconnected and squeezed places are shown as ridges, depressions and deep grooves alternately. The depths range from 5,225 m to 6,206 m, with a maximum total relief of 981 m. The deepest point is located in the central north of the fracture zone.		
专名释义 Reason for Choice of Name	星河断裂带呈弧状弯曲，与东北部的银河断裂带遥相呼应，银河又称“星河”，故名。 Xinghe Fracture Zone is curved in an arc shape, distantly echoing the Yinhe (Chinese name for the Milky Way) Fracture Zone in the northeast. The Yinhe is also called Xinghe (Chinese name for the Star River), hence the feature is named Xinghe Fracture Zone.		

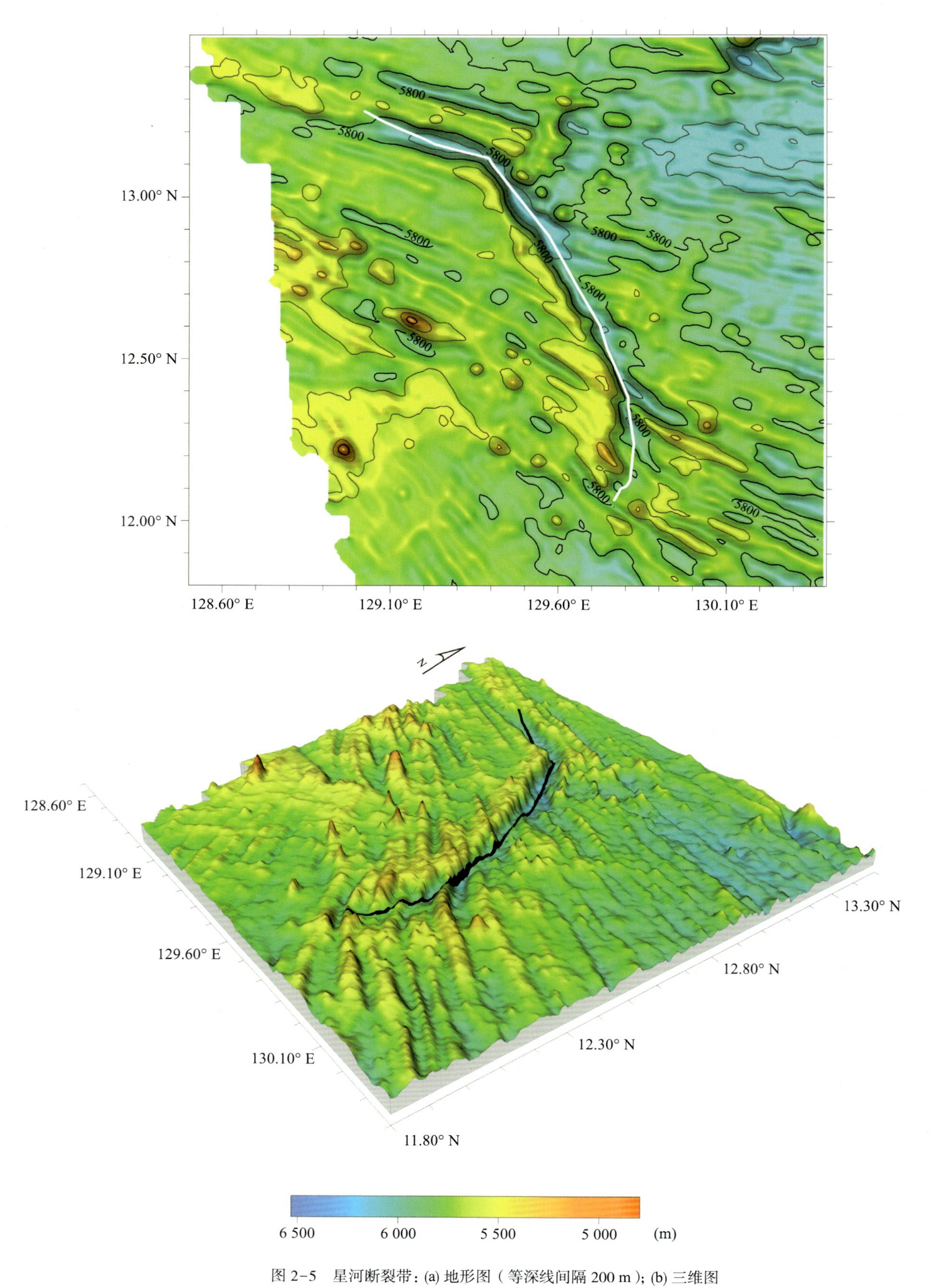

图 2-5　星河断裂带：(a) 地形图（等深线间隔 200 m）；(b) 三维图

Fig.2-5　Xinghe Fracture Zone: (a) Bathymetric map (Contours are in 200 m); (b) 3-D topographic map

2.2.5 弧矢断裂带 Hushi Fracture Zone

<table>
<tr><td>中文名称
Chinese Name</td><td colspan="3">Húshǐ Duànlièdài
弧矢断裂带</td></tr>
<tr><td>英文名称
English Name</td><td colspan="3">Hushi Fracture Zone</td></tr>
<tr><td>地理区域 / Location</td><td colspan="3">西太平洋菲律宾海盆 The West Philippine Basin</td></tr>
<tr><td rowspan="2">特征点坐标
Coordinate</td><td rowspan="2">09°48.6′N 131°33.7′E</td><td>长度 / Length</td><td>280 km</td></tr>
<tr><td>宽度 / Width</td><td>10 km</td></tr>
<tr><td>水深 / Depth</td><td>5 269~6 543 m</td><td>高差 / Total Relief</td><td>1 274 m</td></tr>
<tr><td>发现情况
Discovery Facts</td><td colspan="3">此断裂带于 2011 年“向阳红 14”船在执行航次调查时发现。
The fracture zone was discovered in 2011 during the survey cruise carried out onboard the Chinese R/V Xiang Yang Hong 14.</td></tr>
<tr><td>地形特征
Feature Description</td><td colspan="3">弧矢断裂带位于菲律宾海盆西南部，中央海盆断裂带以南 606 km，呈 NE-SW 向弧状延伸，南北长 280 km，东西宽 10 km。该断裂带受到板块间挤压，地形表现为隆起、间断、挤压合并而交替出现的海脊、海底凹陷、海底深槽地貌，水深范围 5 269~6 543 m，最大高差 1 274 m，最深点位于断裂带内西南部。
Located in the southwest of the West Philippine Basin, Hushi Fracture Zone is located 606 km south of the Central Basin Fracture Rift. Extending in the NE-SW direction like an arc, it is 280 km long from south to north and 10 km wide from west to east. The rising, disconnected and squeezed places of the fracture zone are shown as ridges, depressions and deep grooves alternately. The depths range from 5,269 m to 6,543 m, with a maximum total relief of 1,274m. The deepest point is located in the southwest of the fracture zone.</td></tr>
<tr><td>专名释义
Reason for Choice of Name</td><td colspan="3">弧矢，是我国传统星官体系“三垣二十八宿”南方朱雀第一宿井宿中的星官，位于现代星座划分的大犬座和船尾座。意为“射天狼的弯弓”。该断裂带形似弯弓，故以此命名。
The Hushi, as an asterism in the first constellation Jing Xiu (Well Constellation) of the Nanfang Zhuque in the Three Enclosures and Twenty-Eight Constellations within the traditional Chinese constellation system, is part of the Canis Major and Puppis according to the modern constellation system. The Hushi refers to the bent bow used to shoot the Sirius. As the fracture zone looks like a bent bow, hence the name Hushi.</td></tr>
</table>

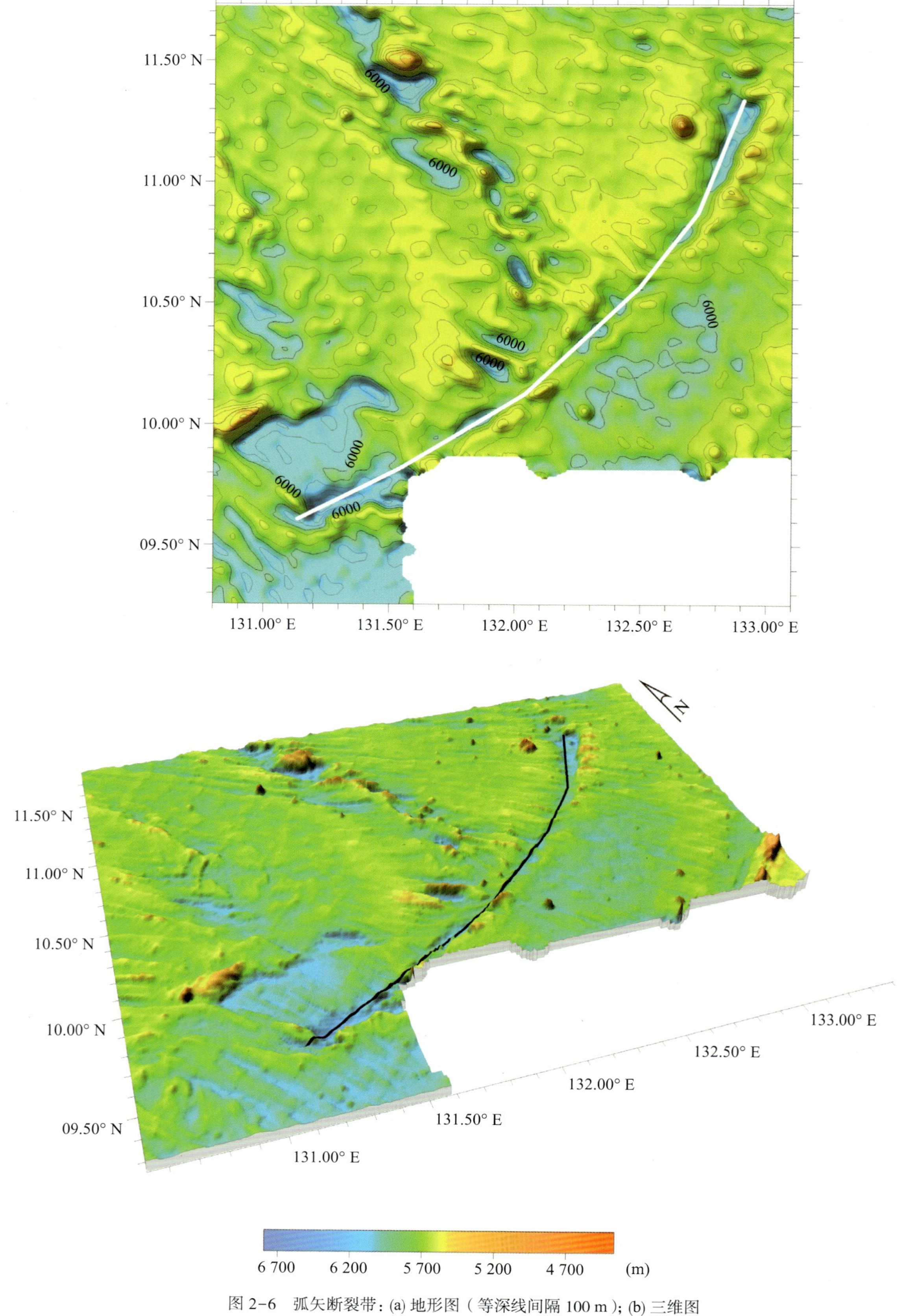

图 2-6 弧矢断裂带：(a) 地形图（等深线间隔 100 m）；(b) 三维图

Fig.2-6 Hushi Fracture Zone: (a) Bathymetric map (Contours are in 100 m); (b) 3-D topographic map

2.2.6 天樽海丘群 Tianzun Hills

中文名称 Chinese Name	Tiānzūn Hǎiqiūqún 天樽海丘群		
英文名称 English Name	Tianzun Hills		
地理区域 / Location	西太平洋菲律宾海盆 The West Philippine Basin		
特征点坐标 Coordinate	15°12.2′N 129°35.3′E 15°06.6′N 129°32.2′E 15°00.7′N 129°27.9′E	长度 / Length 宽度 / Width	36 km 12 km
水深 / Depth	4 744~5 830 m	高差 / Total Relief	1 086 m
发现情况 Discovery Facts	此海丘群于 2006 年“向阳红 14”船、2017 年“向阳红 19”船在执行航次调查时发现。 The hills were discovered in 2006 and 2017 during the survey cruises carried out onboard the Chinese R/V *Xiang Yang Hong 14* and R/V *Xiang Yang Hong 19*, respectively.		
地形特征 Feature Description	天樽海丘群位于菲律宾海盆中部，中央海盆断裂带以南，整体呈 NE-SW 走向，由三座底部相连的海丘组成，三座海丘顶点的相对高差 375 m，南北长 36 km，东西宽 12 km，北侧海丘高于南侧海丘，最高点位于中部海丘，三座海丘中部有凹陷，顶部地形崎岖，南侧海丘有一条向东部延伸的海脊。 Located in the middle of the West Philippine Basin and to the south of Central Basin Fracture Rift. Extending in the NE-SW direction as a whole, they are 36 km long from south to north and 12 km wide from west to east. Consisting of 3 hills which are connected at the bottom, the three shallowest points of the hills have a relative relief of 375 m. The northern hill is a bit higher than the southern hill, and the highest point is located in the middle hill. There are depressions in the middle of the three hills, and it is rugged and uneven at the top of them. In the southern hill, there is a ridge extending to the east.		
专名释义 Reason for Choice of Name	天樽是我国传统星官体系“三垣二十八宿”南方朱雀第一宿井宿中的星官，位于现代星座划分的双子座。天樽意为天上的酒杯。天樽星官有三颗星，位于北河星官附近，就像接着河水的酒樽。该海丘群所包含海丘的个数与天樽星官所包含星星的个数相同，且位于北河断裂带附近，故以此命名。 The name originates from the Tianzun Asterism, which belongs to the first constellation Jing Xiu (Well Constellation) of the Nanfang Zhuque in the Three Enclosures and Twenty-Eight Constellations within the traditional Chinese constellation system. Tianzun Asterism is part of Gemini according to the modern constellation system. The Tianzun refers to the wine glass in the sky. Located near Beihe Asterism, the Tianzun Asterism consists of 3 stars, and looks like a wine goblet that catches river water. The number of the hills and the number of the stars of the Tianzun Asterism are the same, and those hills are located near Beihe Fracture Zone. Therefore, the hills are named Tianzun.		

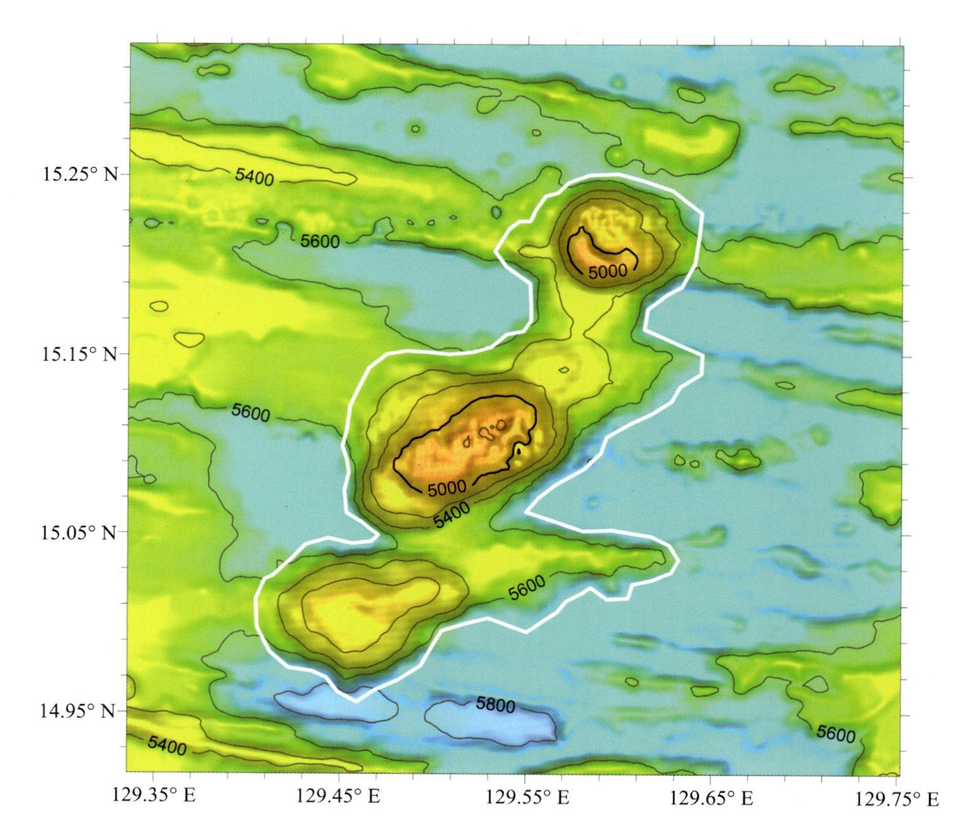

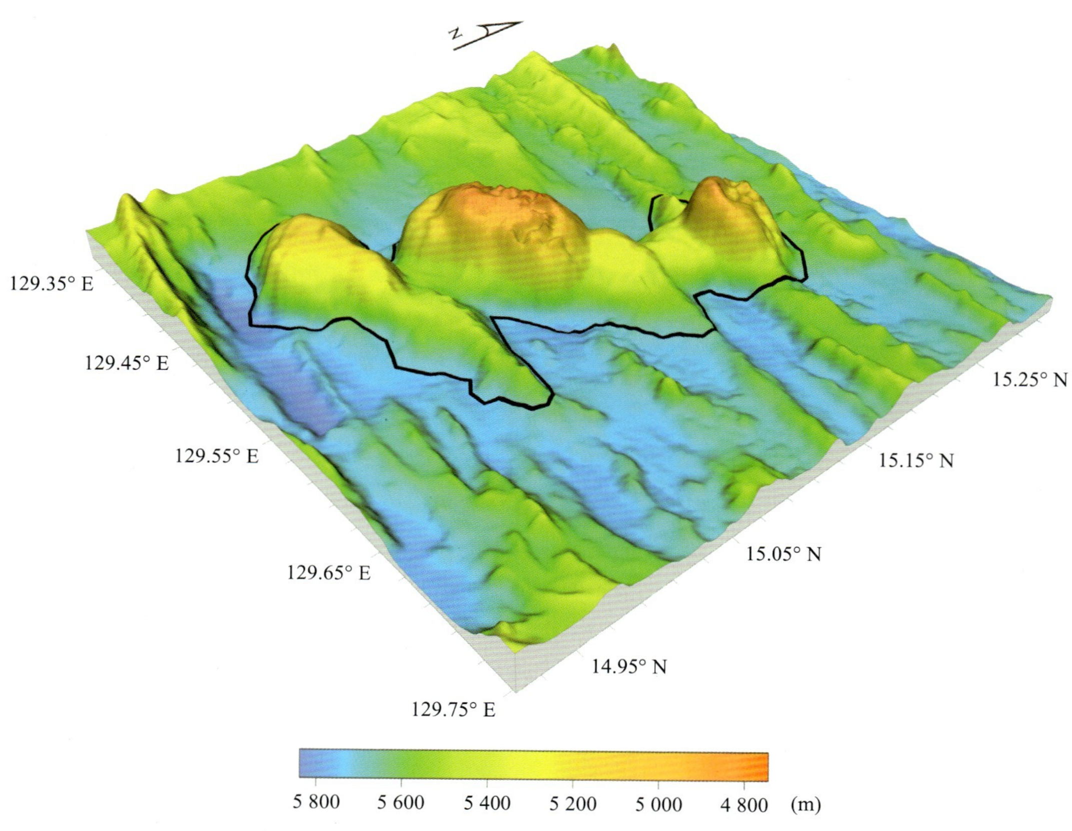

图 2-7 天樽海丘群：(a) 地形图（等深线间隔 200 m）；(b) 三维图

Fig.2-7 Tianzun Hills: (a) Bathymetric map (Contours are in 200 m); (b) 3-D topographic map

2.2.7 积薪海丘 Jixin Hill

中文名称 Chinese Name	Jīxīn Hǎiqiū 积薪海丘		
英文名称 English Name	Jixin Hill		
地理区域 / Location	西太平洋菲律宾海盆 The West Philippine Basin		
特征点坐标 Coordinate	13°56.4′N 130°58.0′E	长度 / Length 宽度 / Width	23 km 11 km
水深 / Depth	4 844~6 281 m	高差 / Total Relief	1 437 m
发现情况 Discovery Facts	此海丘于 2017 年“向阳红 19”船和“向阳红 06”船在执行航次调查时发现。 The hill was discovered in 2017 during the survey cruises carried out onboard the Chinese R/V *Xiang Yang Hong 19* and R/V *Xiang Yang Hong 06*.		
地形特征 Feature Description	积薪海丘位于菲律宾海盆中部，中央海盆断裂带以南，整体呈 E-W 条带走向，长 23 km，宽 11 km，南北陡，东西缓，南北两侧水深急剧增大，突显出该海丘与周围地形的差异。 Located in the middle of the West Philippine Basin and to the south of Central Basin Fracture Rift. Extending in the E-W direction as a whole, it is about 23 km long and 11 km wide. It is steep in the south and the north, and gentle in the west and east. The depth of the southern side and northern side of the hill increases dramatically, which highlights the topographic differences between the hill and its surroundings.		
专名释义 Reason for Choice of Name	名称取自我国传统星官体系“三垣二十八宿”南方朱雀第一宿井宿中的积薪星官，该星官有一颗星，位于南河和北河星官之间，属于现代星座的双子座，意指储存的柴薪，或给厨房官员供应的燃料。该海丘坐落于南河和北河断裂带之间，考虑到地理位置的相似性，故取名积薪。 The name originates from Jixin Asterism, which belongs to the first constellation Jing Xiu (Well constellation) of Nanfang Zhuque in the Three Enclosures and Twenty-Eight Constellations within the traditional Chinese constellation system. Jixin Asterism, located between Nanhe Asterism and Beihe Asterism, has one star and is part of Gemini according to the modern constellation system. Jixin refers to the stored firewood, or the fuels supplied to the kitchen officials. As the hill is located between Beihe Fracture Zone and Nanhe Fracture Zone, it is named Jixin due to the similarity in geographical location.		

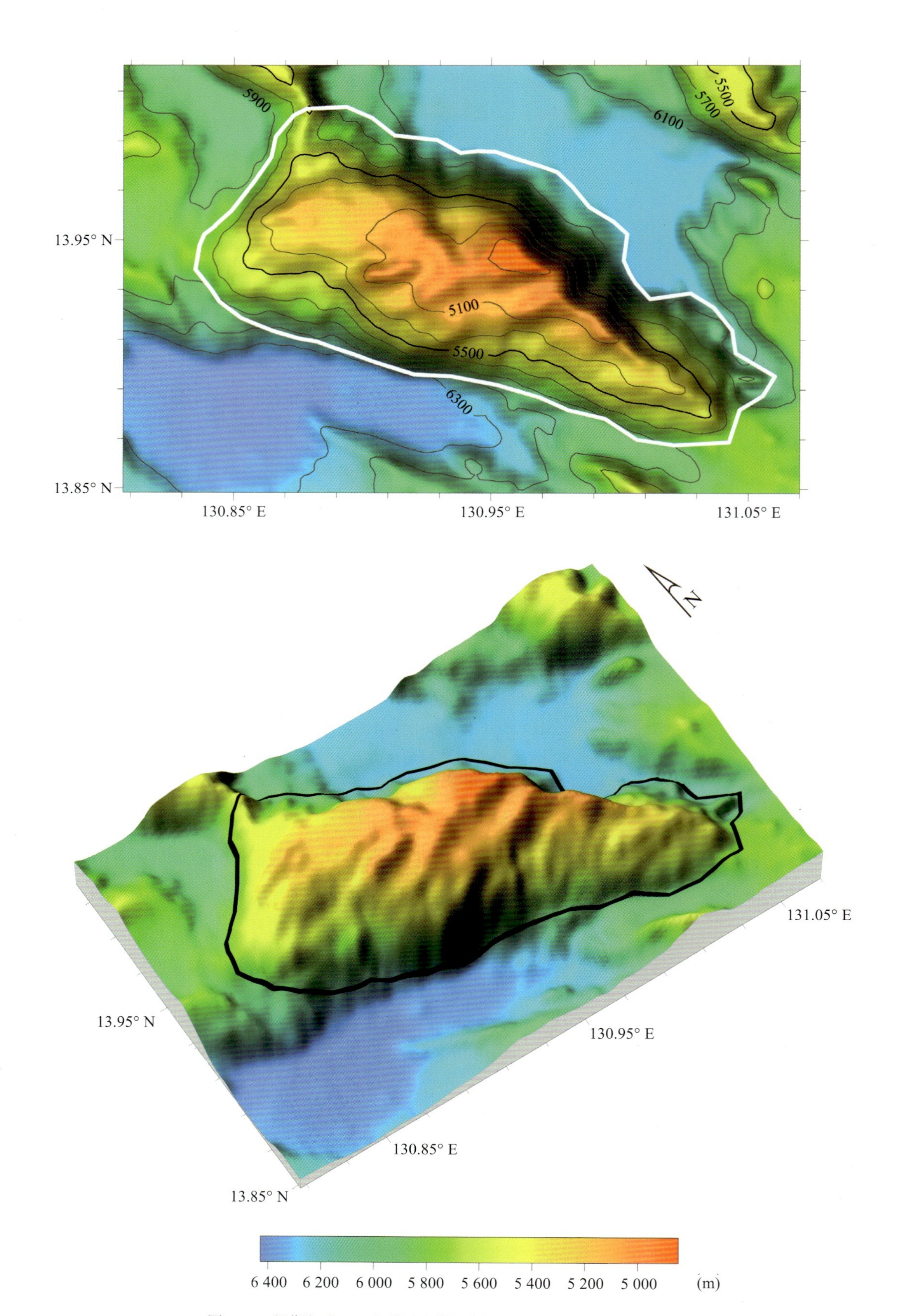

图 2–8　积薪海丘：(a) 地形图（等深线间隔 200 m）；(b) 三维图

Fig.2-8　Jixin Hill: (a) Bathymetric map (Contours are in 200 m); (b) 3-D topographic map

2.2.8 阙丘海丘 Queqiu Hill

中文名称 Chinese Name	Quèqiū Hǎiqiū 阙丘海丘		
英文名称 English Name	Queqiu Hill		
地理区域 / Location	西太平洋菲律宾海盆 The West Philippine Basin		
特征点坐标 Coordinate	12°11.6′N 130°44.6′E	长度 / Length 宽度 / Width	21 km 12 km
水深 / Depth	5 058~5 683 m	高差 / Total Relief	625 m
发现情况 Discovery Facts	此海丘于 2017 年“向阳红 06”船在执行航次调查时发现。 The hill was discovered in 2017 during the survey cruise carried out onboard the Chinese R/V *Xiang Yang Hong 06.*		
地形特征 Feature Description	阙丘海丘位于菲律宾海盆中部，中央海盆断裂带南部。该海丘东高西低，东侧山麓陡峭，自东向西地形先下降再抬升，最后过渡至水深约 5 600 m，海丘顶部较为平缓。 Located in the middle of the West Philippine Basin and to the south of Central Basin Fracture Rift. Being high in the east and low in the west, the hill has a steep foot on the east side. From east to west, the hill first descends, then ascends, and finally transitions to a depth of about 5,600 m, with the top being relatively gentle.		
专名释义 Reason for Choice of Name	名称取自我国传统星官体系“三垣二十八宿”南方朱雀第一宿井宿中的阙丘星官，该星官位于南河和弧矢星官之间，属于现代星座的麒麟座，意为宫门外的小山。该海丘坐落于南河和弧矢断裂带之间，考虑到地理位置的相似性，故取名阙丘。 The name originates from Queqiu Asterism, which belongs to the first constellation Jing Xiu (Well Constellation) of Nanfang Zhuque in the Three Enclosures and Twenty-Eight Constellations within the traditional Chinese constellation system. The Queqiu Asterism, located between the Nanhe Asterism and Hushi Asterism, is part of Monoceros according to the modern constellation system. The Queqiu refers to the hill outside the palace gate. As the hill is located between Nanhe Fracture Zone and Hushi Fracture Zone, it is named Queqiu Hill due to the similarity in geographical location.		

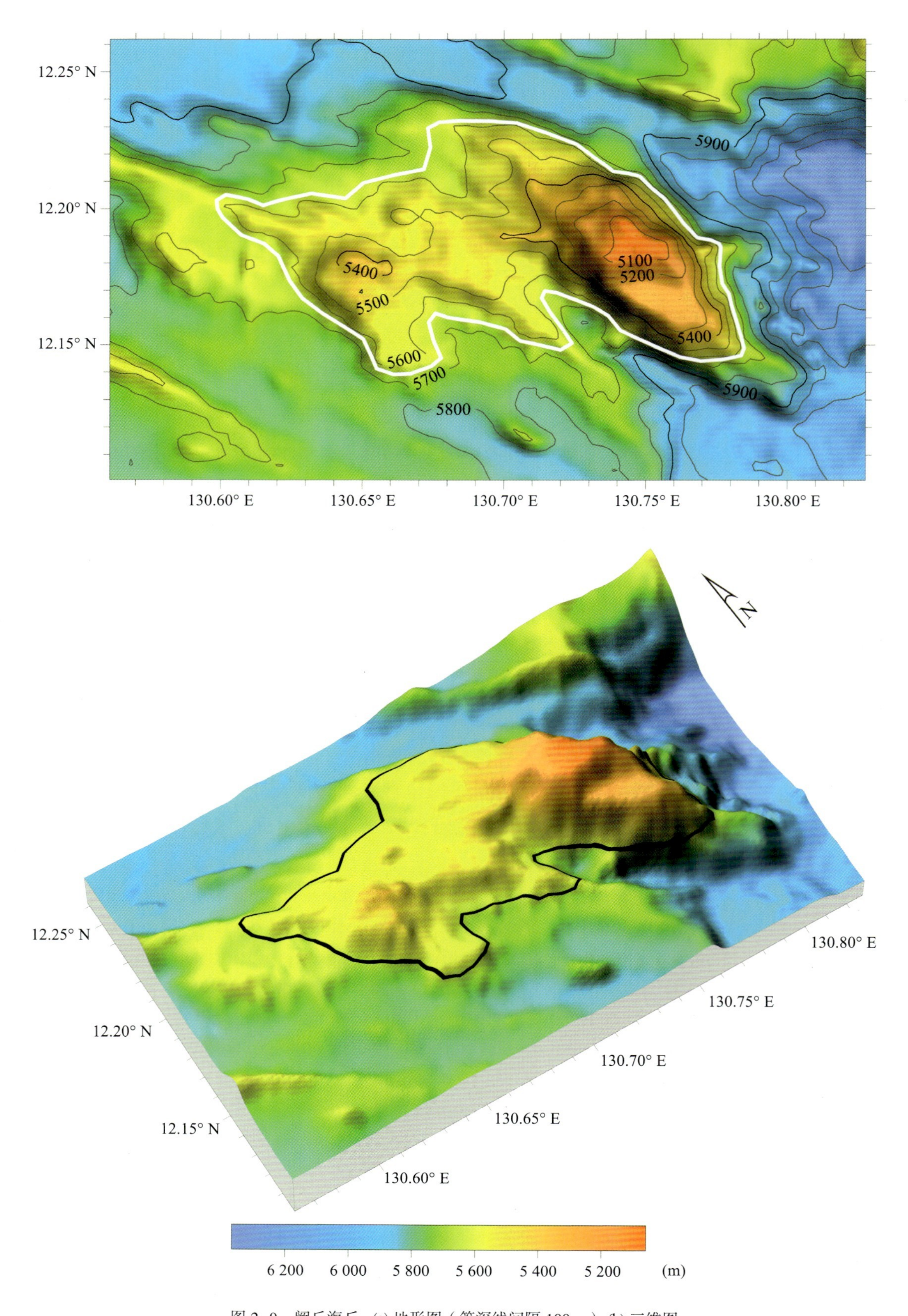

图 2-9　阙丘海丘：(a) 地形图（等深线间隔 100 m）；(b) 三维图

Fig.2-9　Queqiu Hill: (a) Bathymetric map (Contours are in 100 m); (b) 3-D topographic map

2.2.9 浅室海盆 Qianshi Basin

中文名称 Chinese Name	Qiǎnshì Hǎipén 浅室海盆		
英文名称 English Name	Qianshi Basin		
地理区域 / Location	西太平洋菲律宾海盆 The West Philippine Basin		
特征点坐标 Coordinate	16°39.5′N 132°49.4′E	长度 / Length	12 km
		宽度 / Width	6 km
水深 / Depth	5 971~6 308 m	高差 / Total Relief	337 m
发现情况 Discovery Facts	此海盆于 2017 年“向阳红 19”船在执行航次调查时发现。 The basin was discovered in 2017 during the survey cruise carried out onboard the Chinese R/V *Xiang Yang Hong 19.*		
地形特征 Feature Description	浅室海盆位于菲律宾海盆东北部，中央海盆断裂带东北方约 178 km，呈 E-W 向，东宽西窄，最宽约 6 km，东西长 12 km，海盆底部平缓，水深自西北向东北方向逐渐变深。 Located in the northeast of the West Philippine Basin, the basin is about 178 km northeast of Central Basin Fracture Rift. Extending in the E-W direction, it is wide in the east (about 6 km at maximum) and narrow in the west, and measures 12 km from west to east. Being flat at the bottom, the basin becomes increasingly deep from northwest to northeast.		
专名释义 Reason for Choice of Name	该海盆形似一只七棱浅室水母，该水母属于腔肠动物门，在太平洋等热带区域广泛分布，故以浅室命名。 The basin is shaped like an L.multicristata (Moser, or Qileng Qianshi jellyfish in Chinese), which belongs to coelenterata and is widely distributed in tropical ocean such as the Pacific Ocean. This is the very reason why it is called Qianshi Basin.		

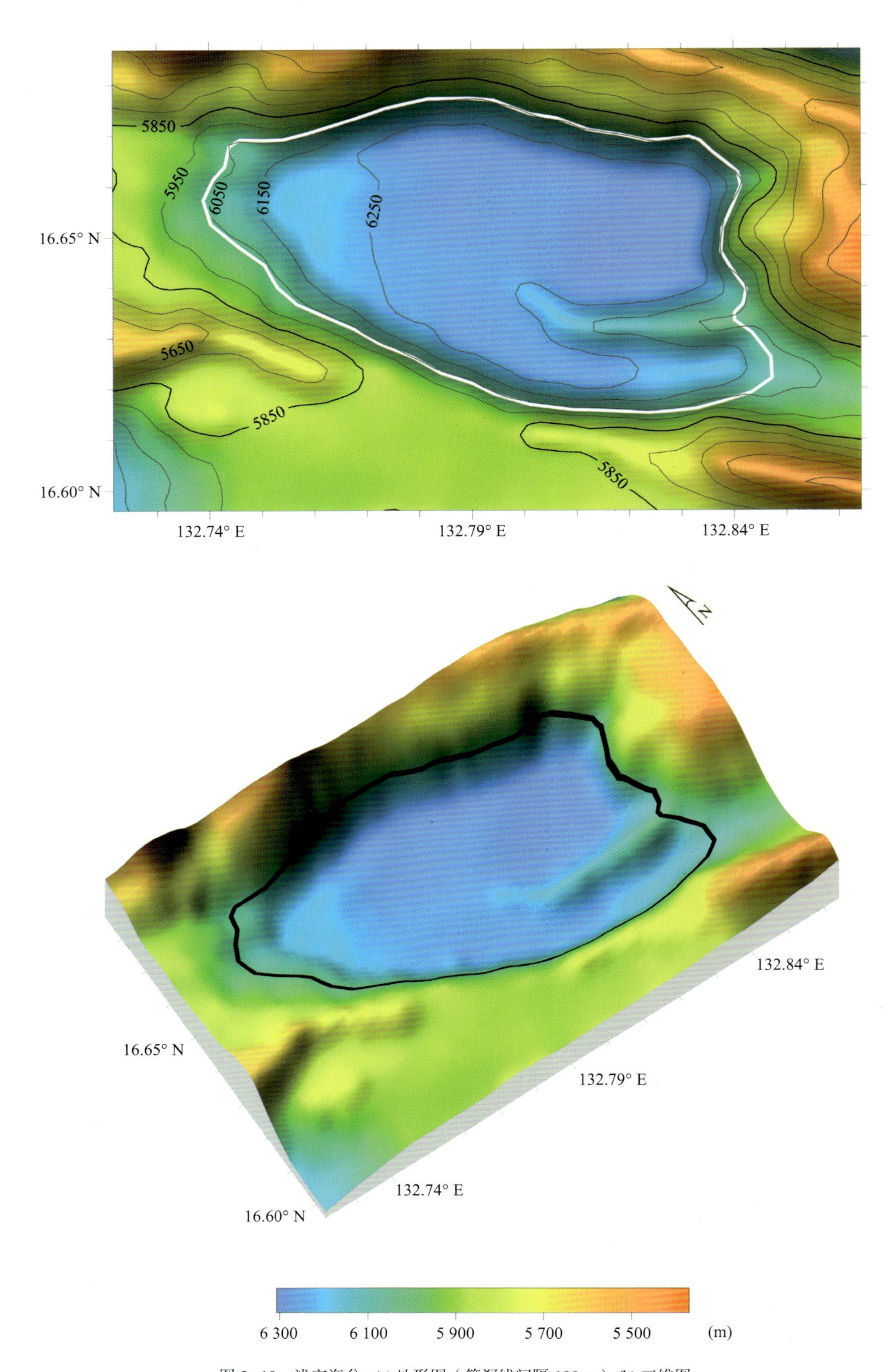

图 2-10 浅室海盆：(a) 地形图（等深线间隔 100 m）；(b) 三维图
Fig.2-10 Qianshi Basin: (a) Bathymetric map (Contours are in 100 m); (b) 3-D topographic map

2.2.10 鱼尾海丘 Yuwei Hill

<table>
<tr><td>中文名称
Chinese Name</td><td colspan="3">Yúwěi Hǎiqiū
鱼尾海丘</td></tr>
<tr><td>英文名称
English Name</td><td colspan="3">Yuwei Hill</td></tr>
<tr><td>地理区域 / Location</td><td colspan="3">西太平洋菲律宾海盆 The West Philippine Basin</td></tr>
<tr><td rowspan="2">特征点坐标
Coordinate</td><td rowspan="2">12°37.7′N 129°09.9′E</td><td>长度 / Length</td><td>15 km</td></tr>
<tr><td>宽度 / Width</td><td>5.5 km</td></tr>
<tr><td>水深 / Depth</td><td>5 031~5 695 m</td><td>高差 / Total Relief</td><td>664 m</td></tr>
<tr><td>发现情况
Discovery Facts</td><td colspan="3">此海丘于 2017 年“向阳红 19”船在执行航次调查时发现。
This hill was discovered in 2017 during the survey cruise carried out onboard the Chinese R/V Xiang Yang Hong 19.</td></tr>
<tr><td>地形特征
Feature Description</td><td colspan="3">鱼尾海丘位于菲律宾海盆中部，中央海盆断裂带以南，整体呈 NW-SE 走向，两端狭窄，中部略宽，南北两侧边界呈弧形，西北侧尾部狭长，南北坡度大，东坡从顶部呈阶梯状逐渐加深。
Located in the middle of the West Philippine Basin, and to the south of Central Basin Fracture Rift. Extending in the NW-SE direction as a whole, it is narrow at both ends and slightly wider in the middle. The hill has arc-shaped borders on both the southern side and northern side, a long and narrow tail on the northwestern side, steep slopes in the south and north, and an eastern slope that is becoming increasingly deep from the top in the ladder pattern.</td></tr>
<tr><td>专名释义
Reason for Choice of Name</td><td colspan="3">该海丘形如俏丽的鱼尾，摆动游弋于深邃的菲律宾海，故以鱼尾命名。
The hill looks like a pretty fish tail (Yuwei in Chinese) protruding from the sea surface and swimming across the deep Philippine Sea, hence the feature is named Yuwei Hill.</td></tr>
</table>

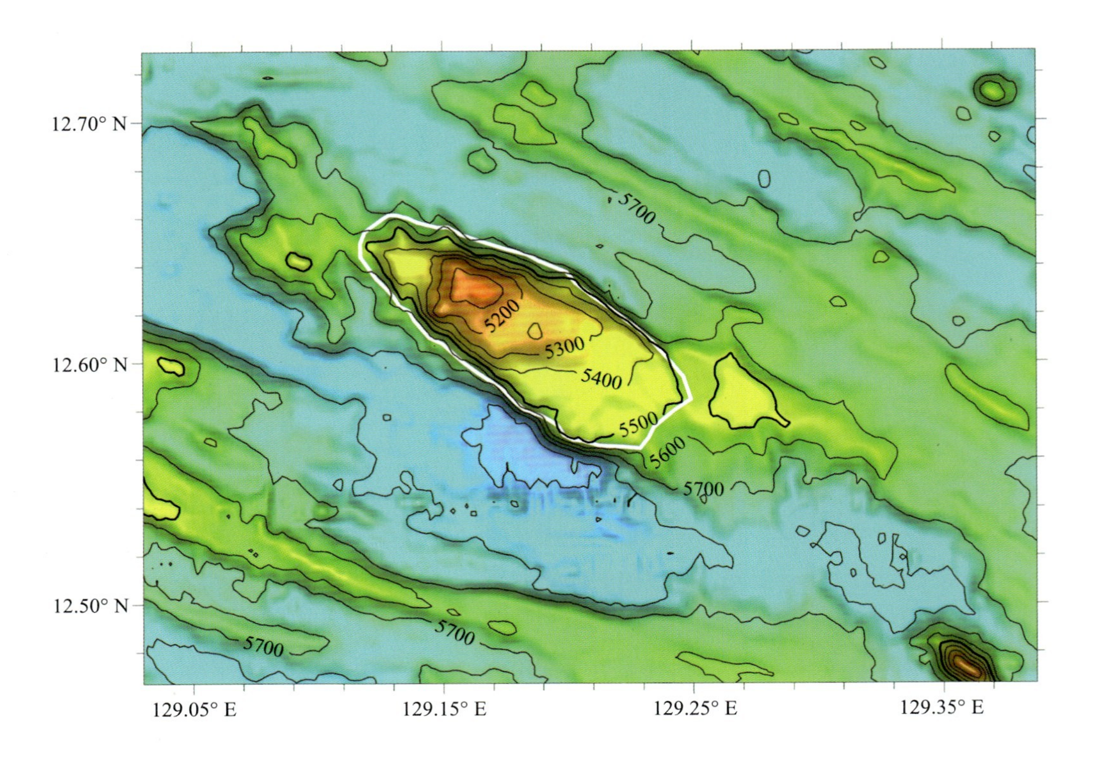

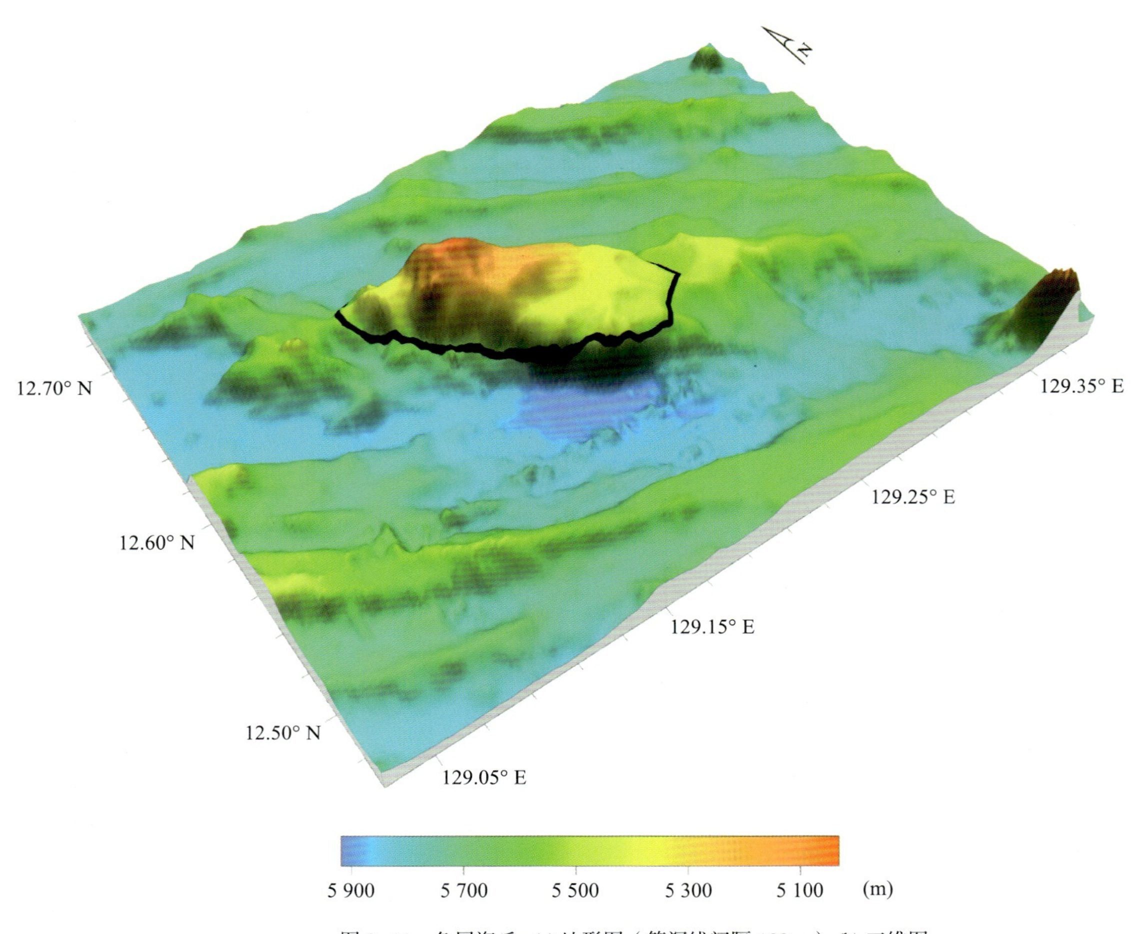

图 2-11 鱼尾海丘：(a) 地形图（等深线间隔 100 m）；(b) 三维图

Fig.2-11 Yuwei Hill: (a) Bathymetric map (Contours are in 100 m); (b) 3-D topographic map

2.2.11 半叶海山 Banye Seamount

中文名称 Chinese Name	Bànyè Hǎishān 半叶海山		
英文名称 English Name	Banye Seamount		
地理区域 / Location	西太平洋菲律宾海盆 The West Philippine Basin		
特征点坐标 Coordinate	09°14.3′N 130°07.9′E	长度 / Length	24 km
		宽度 / Width	7 km
水深 / Depth	4 829~5 832 m	高差 / Total Relief	1 003 m
发现情况 Discovery Facts	此海山于 2017 年"向阳红 19"船在执行航次调查时发现。 The seamount was discovered in 2017 during the survey cruise carried out onboard the Chinese R/V *Xiang Yang Hong 19.*		
地形特征 Feature Description	半叶海山位于菲律宾海盆南部，中央海盆断裂带以南，整体呈 NW-SE 走向，两端狭长，中部略宽，西部呈弧形，地形平缓，东部陡直呈断崖状地形。 Located in the south of the West Philippine Basin, and to the south of the Central Basin Fracture Rift. Extending in the NW-SE direction as a whole, it is long and narrow at both ends, and slightly wider in the middle. The arc-shaped west is gentle whereas the east is as steep as a precipice.		
专名释义 Reason for Choice of Name	该海山酷似一片从主干叶脉截去右半部分的树叶，故以半叶命名。 Shaped like a leaf with the right half cut off from the main vein, hence the feature is named Banye Seamount (Banye means half a tree leaf in Chinese).		

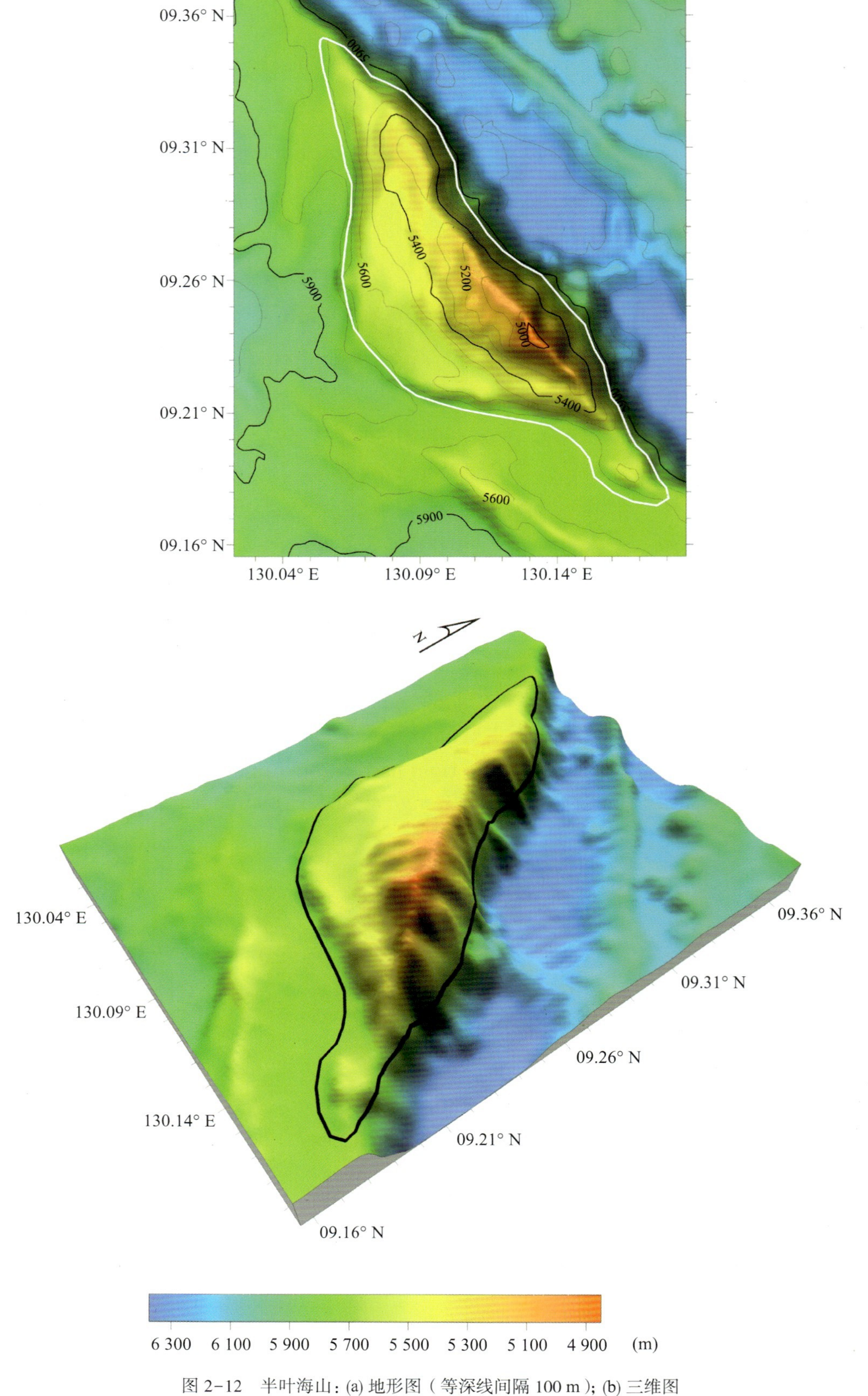

图 2-12　半叶海山：(a) 地形图（等深线间隔 100 m）；(b) 三维图

Fig.2-12　Banye Seamount: (a) Bathymetric map (Contours are in 100 m); (b) 3-D topographic

2.2.12 摇光海山 Yaoguang Seamount

中文名称 Chinese Name	Yáoguāng Hǎishān 摇光海山		
英文名称 English Name	Yaoguang Seamount		
地理区域 / Location	西太平洋菲律宾海盆 The West Philippine Basin		
特征点坐标 Coordinate	18°08.3′N 139°50.0′E	长度 / Length	14 km
		宽度 / Width	13 km
水深 / Depth	3 376~4 668 m	高差 / Total Relief	1 292 m
发现情况 Discovery Facts	此海山于 2018 年“向阳红 10”船在执行航次调查时发现。 The seamount was discovered in 2018 during the survey cruise carried out onboard the Chinese R/V *Xiang Yang Hong 10.*		
地形特征 Feature Description	摇光海山位于帕里西维拉海盆中部，帕里西维拉裂谷以东约 50 km，其形体浑圆，西北坡缓，东南坡陡，该海山顶部有一近似圆形凹陷。 Located in the middle of the Parece Vela Basin, the seamount is about 50 km east of the Parece Vela Rift. Being rounded in shape, it has a gentle slope in the northwest and a steep slope in the southeast. At the top of the seamount, there is a generally rounded depression.		
专名释义 Reason for Choice of Name	名称取自我国传统星官体系“三垣二十八宿”紫薇垣星官中的北斗七星，北斗七星包括三颗斗柄星和四颗斗魁星，该海山和南部其他六个海山依北斗七星分布方位命名，其中，摇光位于北斗七星斗柄的最末端（第七颗星）。 The name originates from the Big Dipper of the Ziwei Enclosure Constellation of the Three Enclosures and Twenty-Eight Constellations within the traditional Chinese constellation system. The Big Dipper consists of three handle stars and four bowl stars. This feature and other six seamounts in the south are named according to the star distribution of the Big Dipper. Among them, Yaoguang is at the very end (the seventh star) of the handle of the Big Dipper.		

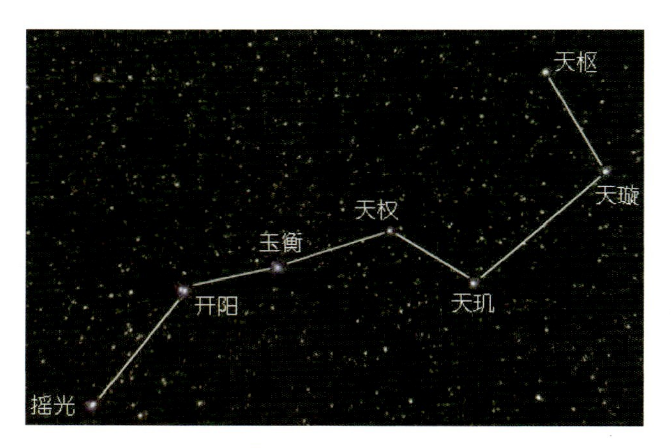

图 2-13　北斗七星
Fig.2-13　The Big Dipper

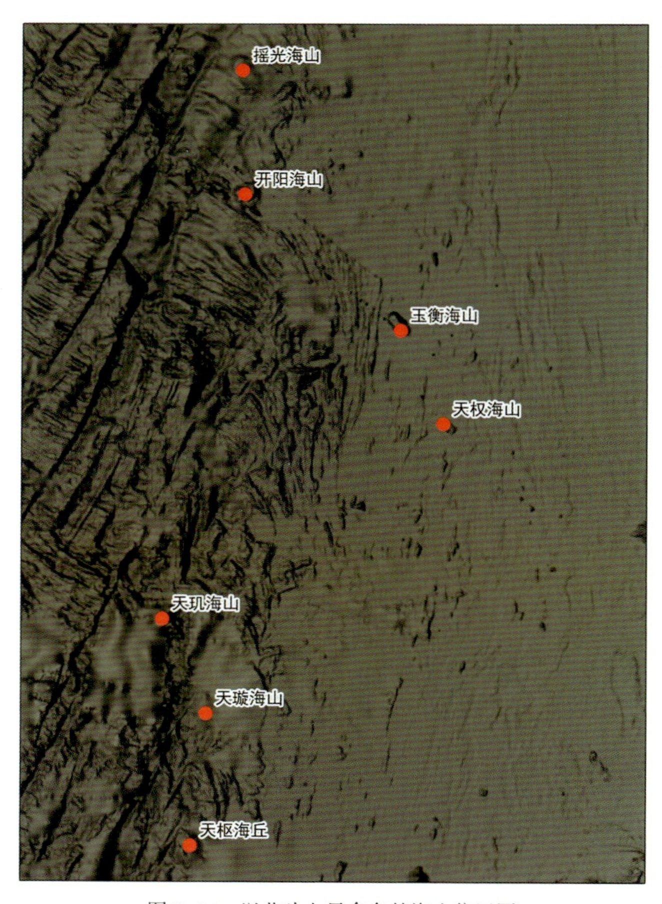

图 2-14　以北斗七星命名的海山位置图
Fig.2-14　Map of the seamounts named according to the Big Dipper star

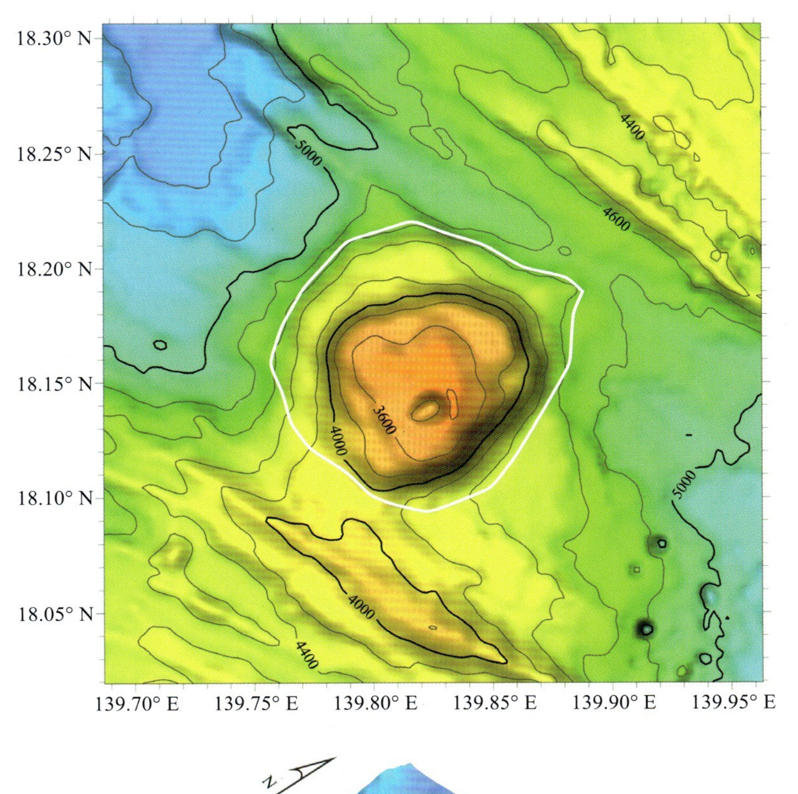

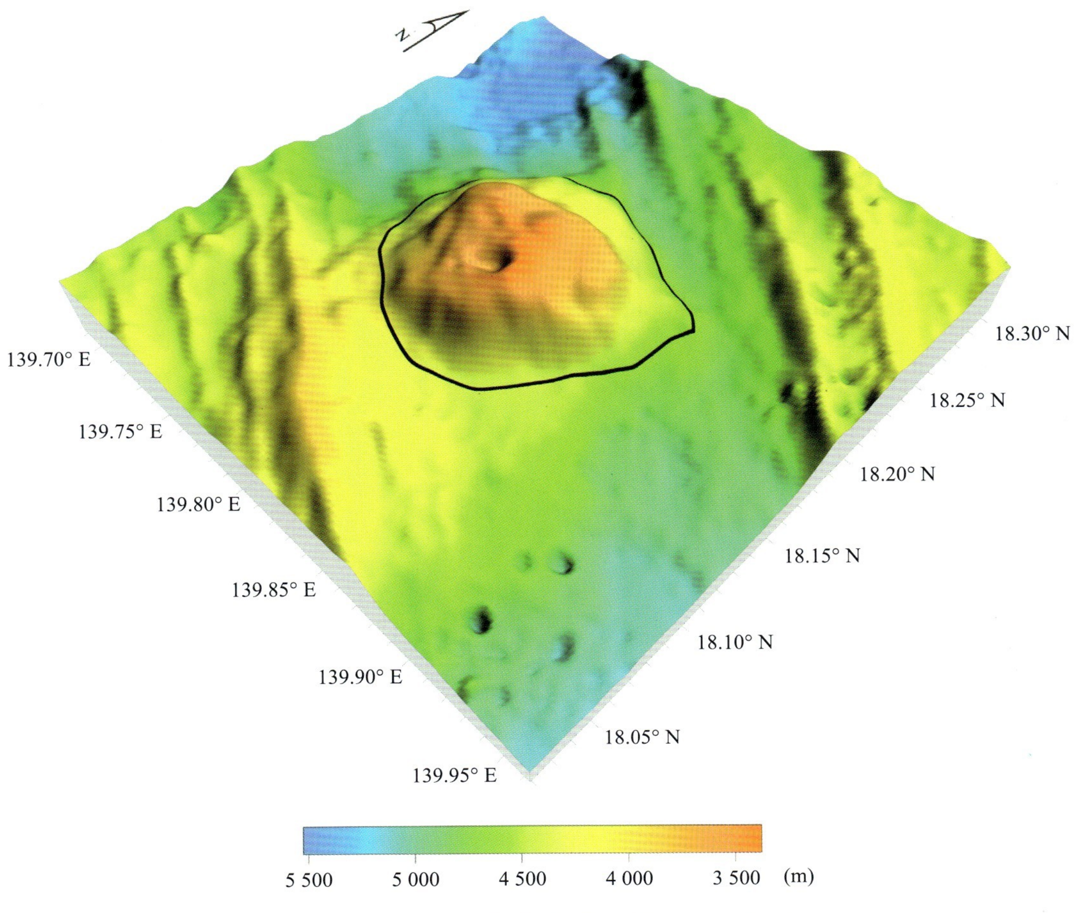

图 2-15 摇光海山：(a) 地形图（等深线间隔 200 m）；(b) 三维图

Fig.2-15 Yaoguang Seamount: (a) Bathymetric map (Contours are in 200 m); (b 3-D topographic map

2.2.13 开阳海山 Kaiyang Seamount

<table>
<tr><td>中文名称
Chinese Name</td><td colspan="4">Kāiyáng Hǎishān
开阳海山</td></tr>
<tr><td>英文名称
English Name</td><td colspan="4">Kaiyang Seamount</td></tr>
<tr><td>地理区域 / Location</td><td colspan="4">西太平洋菲律宾海盆 The West Philippine Basin</td></tr>
<tr><td rowspan="2">特征点坐标
Coordinate</td><td rowspan="2">17°25.6′N 139°50.9′E</td><td>长度 / Length</td><td colspan="2">22.7 km</td></tr>
<tr><td>宽度 / Width</td><td colspan="2">19.6 km</td></tr>
<tr><td>水深 / Depth</td><td>3 162~4 873 m</td><td>高差 / Total Relief</td><td colspan="2">1 711 m</td></tr>
<tr><td>发现情况
Discovery Facts</td><td colspan="4">此海山于 2018 年“向阳红 10”船在执行航次调查时发现。
The seamount was discovered in 2018 during the survey cruise carried out onboard the Chinese R/V Xiang Yang Hong 10.</td></tr>
<tr><td>地形特征
Feature Description</td><td colspan="4">开阳海山位于帕里西维拉海盆中部，帕里西维拉裂谷以东约 40 km，摇光海山以北 76 km。该海山顶部平坦，逐步呈平台状向四周下降。
Located in the middle of the Parece Vela Basin, the seamount is about 40 km east of the Parece Vela Rift, and 76 km north of Yaoguang Seamount. The seamount is flat at the top, and gradually descends to the surroundings in a platform shape.</td></tr>
<tr><td>专名释义
Reason for Choice of Name</td><td colspan="4">名称取自我国传统星官体系“三垣二十八宿”紫薇垣星官中北斗七星的斗柄，该海山和其他六个海山依北斗七星分布方位命名，开阳是北斗七星斗柄的第二颗星。
The name originates from the handle of the Big Dipper of the Ziwei Enclosure Constellation of the Three Enclosures and Twenty-Eight Constellations in the traditional Chinese constellation system. This feature and other six seamounts are named according to star distribution of the Big Dipper. Among them, Kaiyang is the second star of the handle of the Big Dipper.</td></tr>
</table>

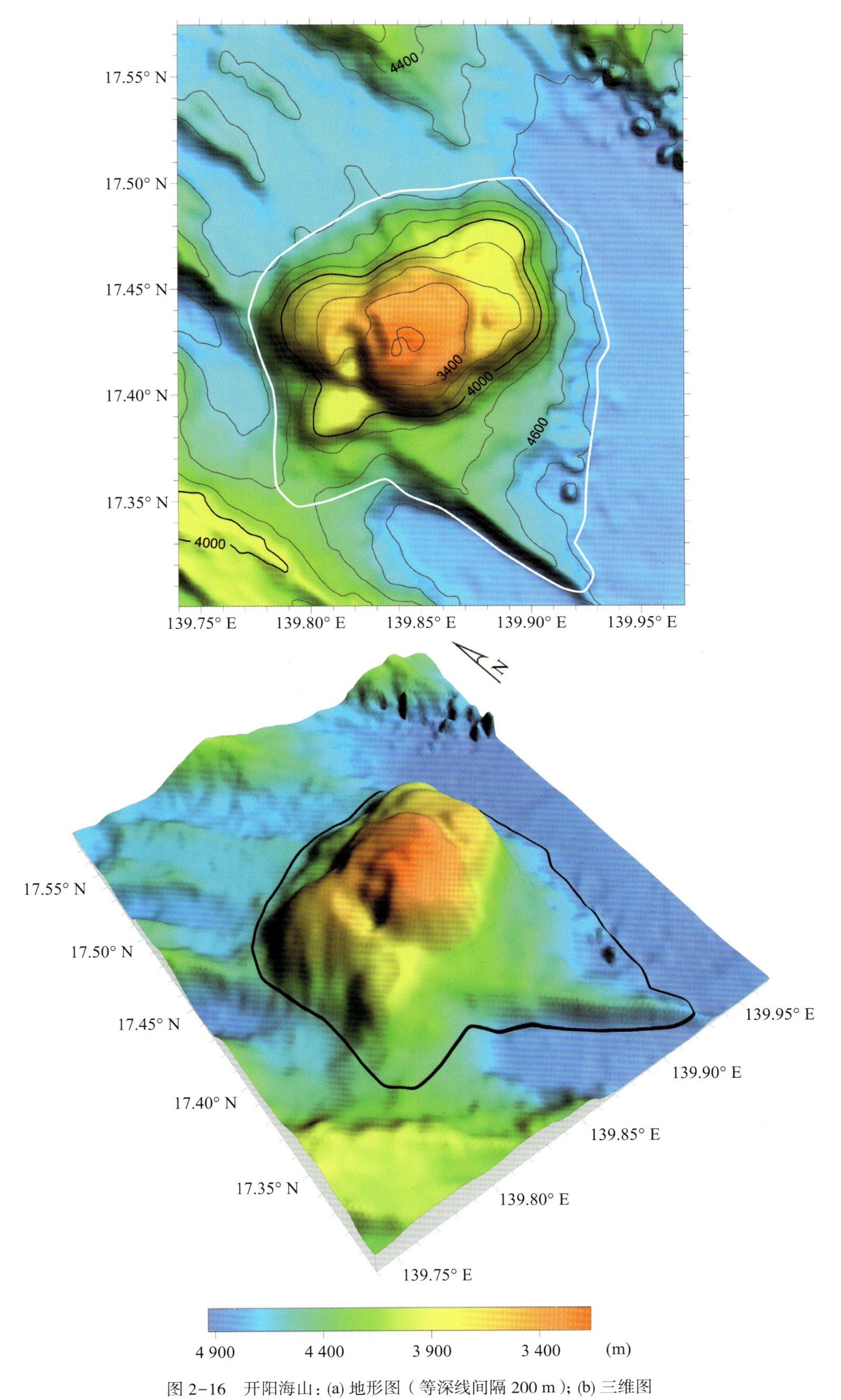

图 2–16 开阳海山：(a) 地形图（等深线间隔 200 m）；(b) 三维图

Fig.2-16 Kaiyang Seamount: (a) Bathymetric map (Contours are in 200 m); (b) 3-D topographic map

2.2.14 玉衡海山 Yuheng Seamount

中文名称 Chinese Name	Yùhéng Hǎishān 玉衡海山		
英文名称 English Name	Yuheng Seamount		
地理区域 / Location	西太平洋菲律宾海盆 The West Philippine Basin		
特征点坐标 Coordinate	16°39.8′N 140°42.5′E	长度 / Length	17.2 km
		宽度 / Width	9.8 km
水深 / Depth	3 615~4 809 m	高差 / Total Relief	1 194 m
发现情况 Discovery Facts	此海山于 2017 年“向阳红 01”船在执行航次调查时发现。 The seamount was discovered in 2017 during the survey cruise carried out onboard the Chinese R/V *Xiang Yang Hong 01*.		
地形特征 Feature Description	玉衡海山位于帕里西维拉海盆中部，帕里西维拉裂谷以东约 110 km，开阳海山东南约 130 km，整体呈 NW-SE 走向，四周坡度较陡，海山顶部西北深约 4 809 m，东南浅约 3 615 m。 Located in the middle of the Parece Vela Basin, the seamount is about 110 km east of the Parece Vela Rift, and about 130 km southeast of Kaiyang Seamount. Extending in the NW-SE direction as a whole, it has relatively steep slopes around. At the top of the seamount, it is about 4,000 m deep in the northwest and about 3,650 m deep in the southeast.		
专名释义 Reason for Choice of Name	名称取自我国传统星官体系“三垣二十八宿”紫薇垣星官中北斗七星的斗柄，该海山和其他六个海山依北斗七星分布方位命名，玉衡是北斗七星斗柄的第一颗星。 The name originates from the handle of the Big Dipper of the Ziwei Enclosure Constellation of the Three Enclosures and Twenty-Eight Constellations in the traditional Chinese constellation system. This feature and other six seamounts are named according to the star distribution of the Big Dipper. Among them, the Yuheng is the first star of the handle of the Big Dipper.		

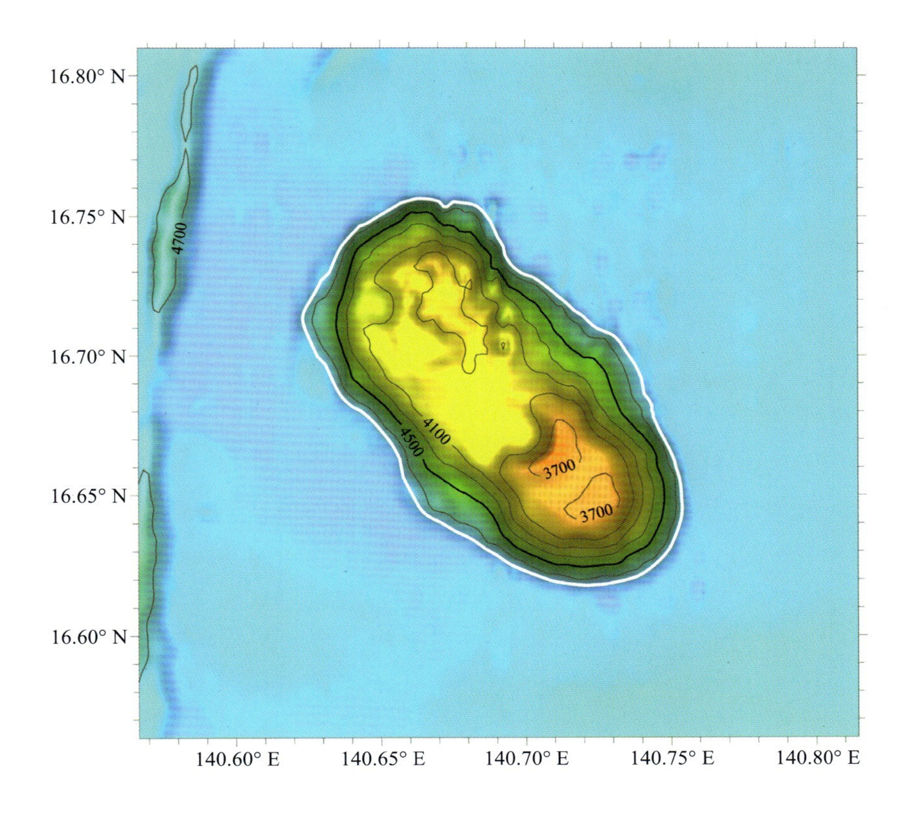

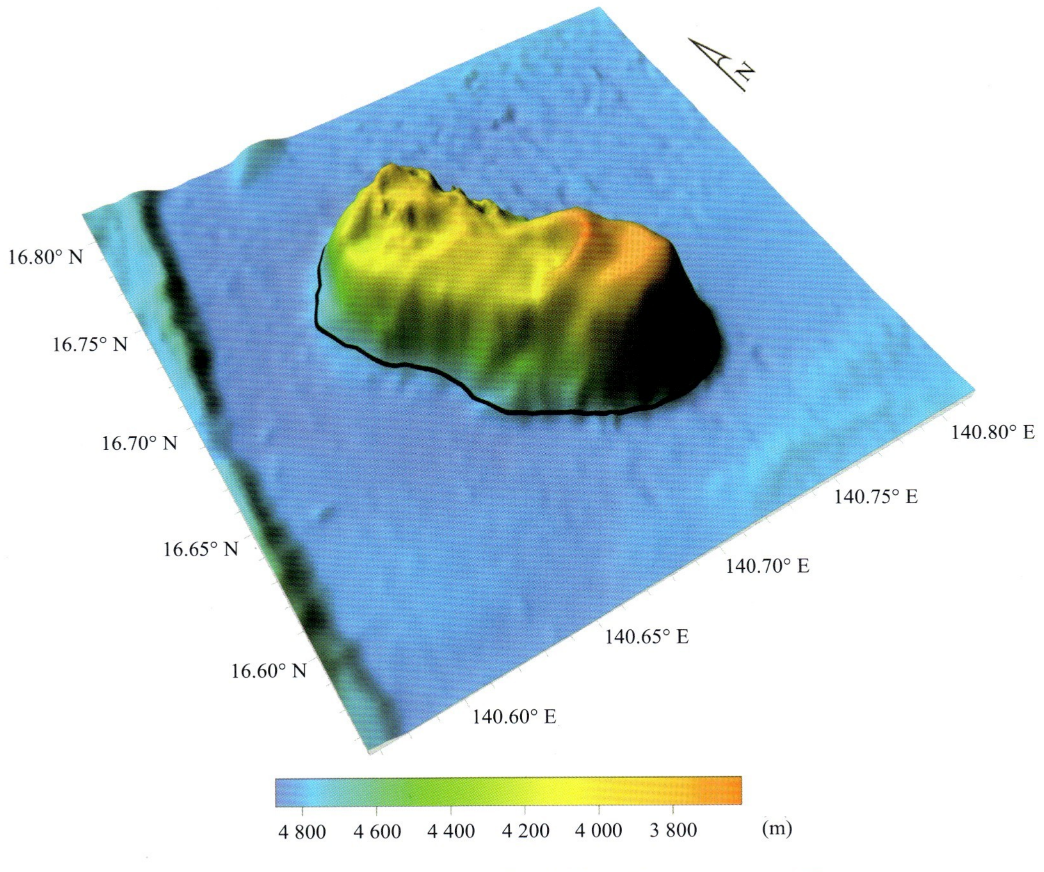

图 2-17 玉衡海山：(a) 地形图（等深线间隔 200 m）；(b) 三维图

Fig.2-17 Yuheng Seamount: (a) Bathymetric map (Contours are in 200 m); (b) 3-D topographic map

2.2.15 天权海山 Tianquan Seamount

中文名称 Chinese Name	Tiānquán Hǎishān 天权海山		
英文名称 English Name	Tianquan Seamount		
地理区域 / Location	西太平洋菲律宾海盆 The West Philippine Basin		
特征点坐标 Coordinate	16°06.8′N 140°56.7′E	长度 / Length	12 km
		宽度 / Width	6.5 km
水深 / Depth	3 771~4 776 m	高差 / Total Relief	1 005 m
发现情况 Discovery Facts	此海山于 2017 年“向阳红 01”船在执行航次调查时发现。 The seamount was discovered in 2017 during the survey cruise carried out onboard the Chinese R/V *Xiang Yang Hong 01*.		
地形特征 Feature Description	天权海山位于帕里西维拉海盆中部，帕里西维拉裂谷以东 155 km，玉衡海山东南 67 km，整体呈 NW–SE 走向，东南部坡度平缓，西北部坡度较陡。 Located in the middle of the Parece Vela Basin, the seamount is 155 km east of the Parece Vela Rift, and 67 km southeast of Yuheng Seamount. Extending in the NW-SE direction as a whole, it has a gentle slope in the southeast and a steep slope in the northwest.		
专名释义 Reason for Choice of Name	名称取自我国传统星官体系“三垣二十八宿”紫薇垣星官中北斗七星勺头的四颗斗魁星，该海山和其他六个海山依北斗七星分布方位命名，天权是北斗七星斗魁星中的第四颗星。 The name originates from the four bowl stars of the Big Dipper of the Ziwei Enclosure Constellation of the Three Enclosures and Twenty-Eight Constellations in the traditional Chinese constellation system. This feature and other six seamounts are named according to the star distribution of the Big Dipper. Among them, Tianquan is the fourth star.		

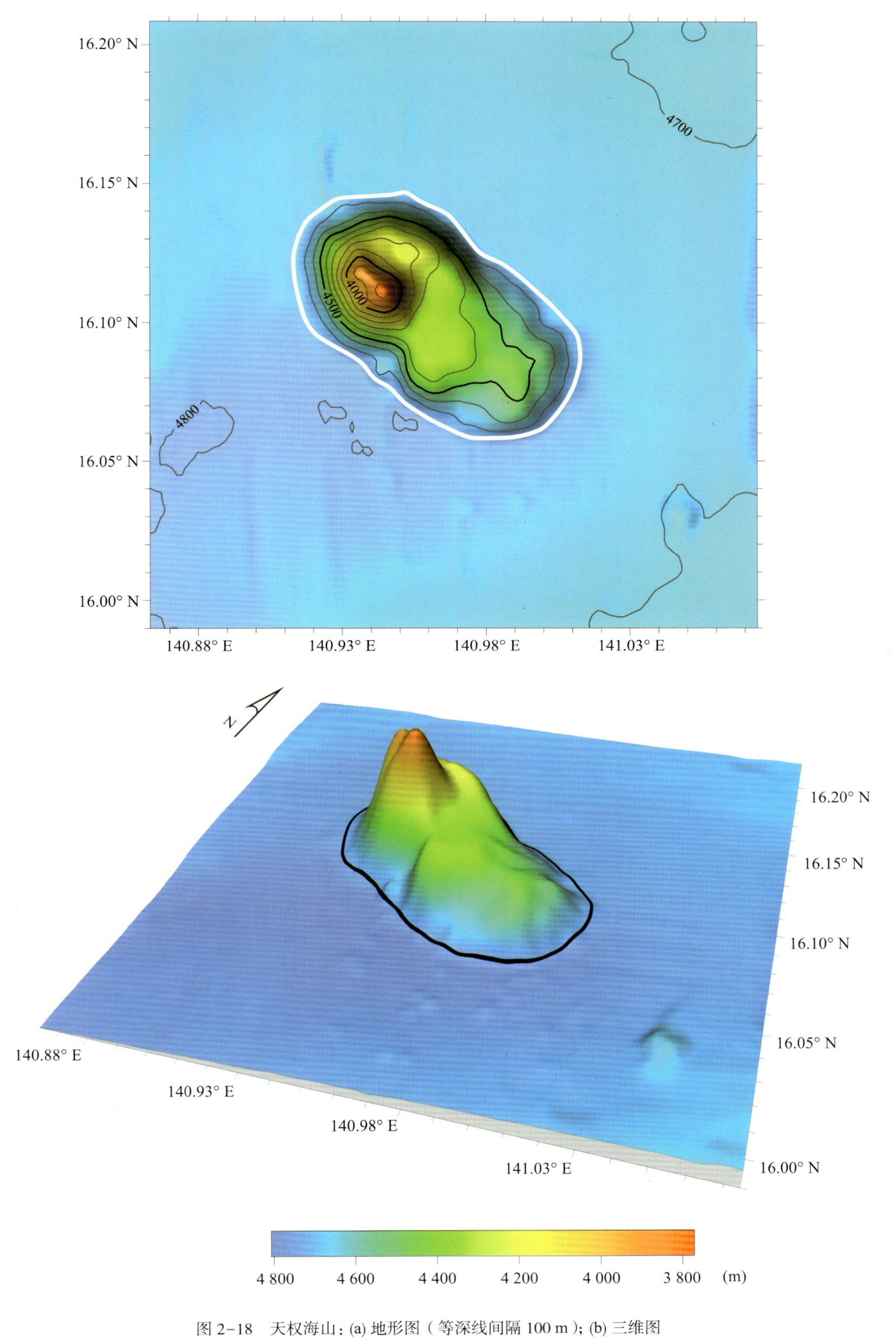

图 2-18 天权海山：(a) 地形图（等深线间隔 100 m）；(b) 三维图

Fig.2-18 Tianquan Seamount: (a) Bathymetric map (Contours are in 100 m); (b) 3-D topographic map

2.2.16 天玑海山 Tianji Seamount

中文名称 Chinese Name	Tiānjī Hǎishān 天玑海山		
英文名称 English Name	Tianji Seamount		
地理区域 / Location	西太平洋菲律宾海盆 The West Philippine Basin		
特征点坐标 Coordinate	14°59.9′N 139°23.9′E	长度 / Length	29 km
		宽度 / Width	18 km
水深 / Depth	2 518~5 277 m	高差 / Total Relief	2 759 m
发现情况 Discovery Facts	此海山于 2014 年“向阳红 14”船在执行航次调查时发现。 The seamount was discovered in 2014 during the survey cruise carried out onboard the Chinese R/V *Xiang Yang Hong 14*.		
地形特征 Feature Description	天玑海山位于帕里西维拉海盆南部，帕里西维拉裂谷以南 120 km，天权海山西南 215 km，整体呈 S-N 走向，有两条长 9 km 的山脊向南延伸，该海山四周崎岖不平。 Located in the south of the Parece Vela Basin, the seamount is 120 km south of the Parece Vela Rift, and 215 km southwest of Tianquan Seamount. Extending in the S-N direction as a whole, the seamount has two 9 km-long ridges extending to the south, and is rugged and uneven in its surroundings.		
专名释义 Reason for Choice of Name	名称取自我国传统星官体系“三垣二十八宿”紫薇垣星官中北斗七星的勺头的四颗斗魁星，该海山和其他六个海山依北斗七星分布方位命名，天玑是北斗七星斗魁星中的第三颗星。 The name originates from the four bowl stars of the Big Dipper of the Ziwei Enclosure Constellation of the Three Enclosures and Twenty-Eight Constellations in the traditional Chinese constellation system. This feature and other six seamounts are named according to the star distribution of the Big Dipper. Among them, the Tianji is the third star.		

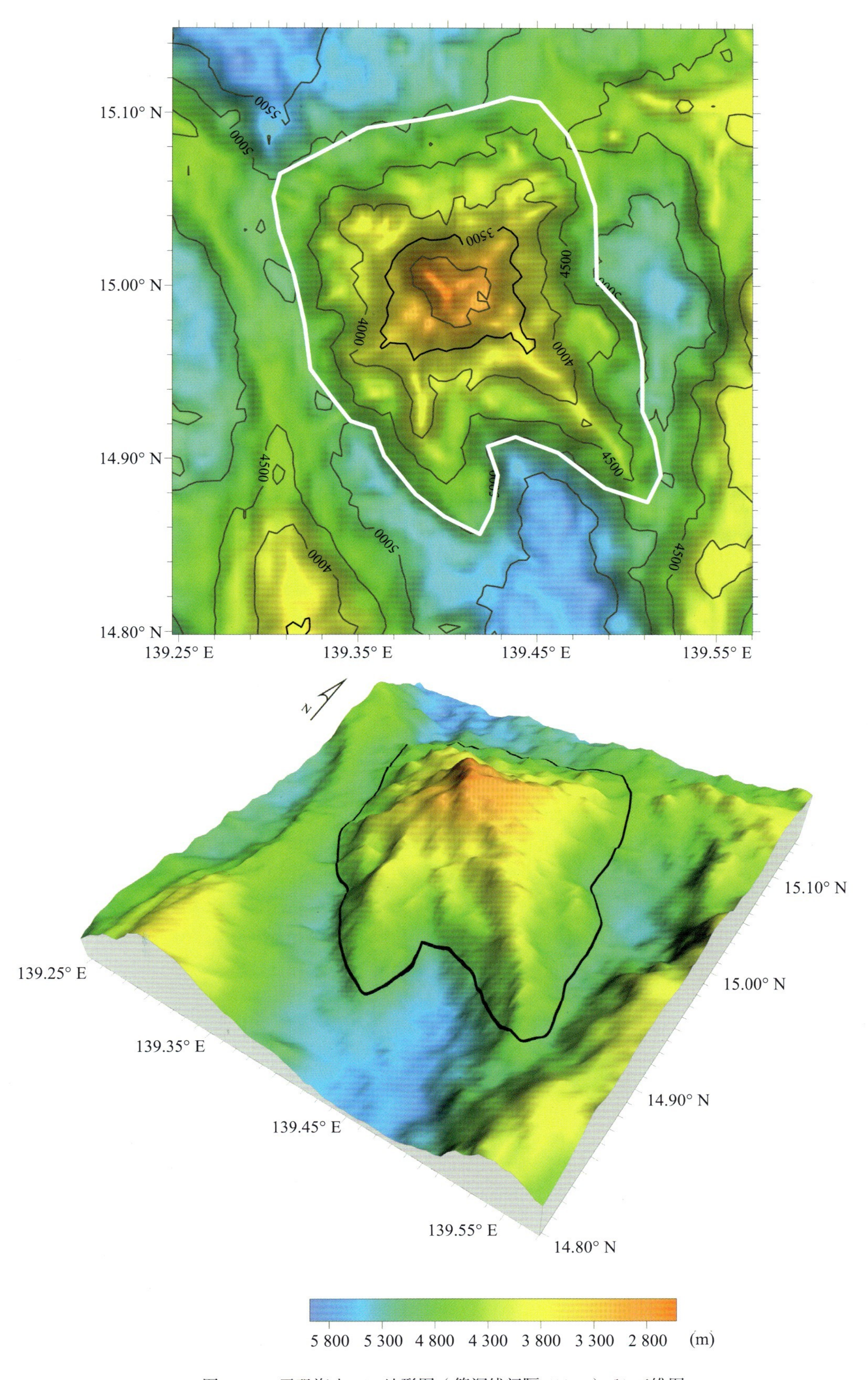

图 2-19 天玑海山：(a) 地形图（等深线间隔 500 m）；(b) 三维图
Fig.2-19 Tianji Seamount: (a) Bathymetric map (Contours are in 500 m); (b) 3-D topographic

2.2.17 天璇海山 Tianxuan Seamount

中文名称 Chinese Name	Tiānxuán Hǎishān 天璇海山		
英文名称 English Name	Tianxuan Seamount		
地理区域 / Location	西太平洋菲律宾海盆 The West Philippine Basin		
特征点坐标 Coordinate	14°28.1′N 139°38.5′E	长度 / Length	22.3 km
		宽度 / Width	18 km
水深 / Depth	3 525~5 077 m	高差 / Total Relief	2 552 m
发现情况 Discovery Facts	此海山于2014年“向阳红14”船在执行航次调查时发现。 The seamount was discovered in 2014 during the survey cruise carried out onboard the Chinese R/V *Xiang Yang Hong 14*.		
地形特征 Feature Description	天璇海山位于帕里西维拉海盆中部，帕里西维拉裂谷以南177 km，天玑海山东南66 km，整体呈等维展布，东坡和南坡平缓，西北坡较陡。 Located in the central part of the Parece Vela Basin, the seamount is 177 km south of the Parece Vela Rift, and 66 km southeast of Tianji Seamount. It exhibits an equidimensional distribution, with gentle slopes on its eastern and southern flanks, while the western flank is relatively steep.		
专名释义 Reason for Choice of Name	名称取自我国传统星官体系“三垣二十八宿”紫薇垣星官中北斗七星的勺头的四颗斗魁星，该海山和其他六个海山依北斗七星分布方位命名，天璇是北斗七星斗魁星中的第二颗星。 The name originates from the four bowl stars of the Big Dipper of the Ziwei Enclosure Constellation of the Three Enclosures and Twenty-Eight Constellations in the traditional Chinese constellation system. This feature and other six seamounts are named according to the star distribution of the Big Dipper. Among them, the Tianxuan is the second star.		

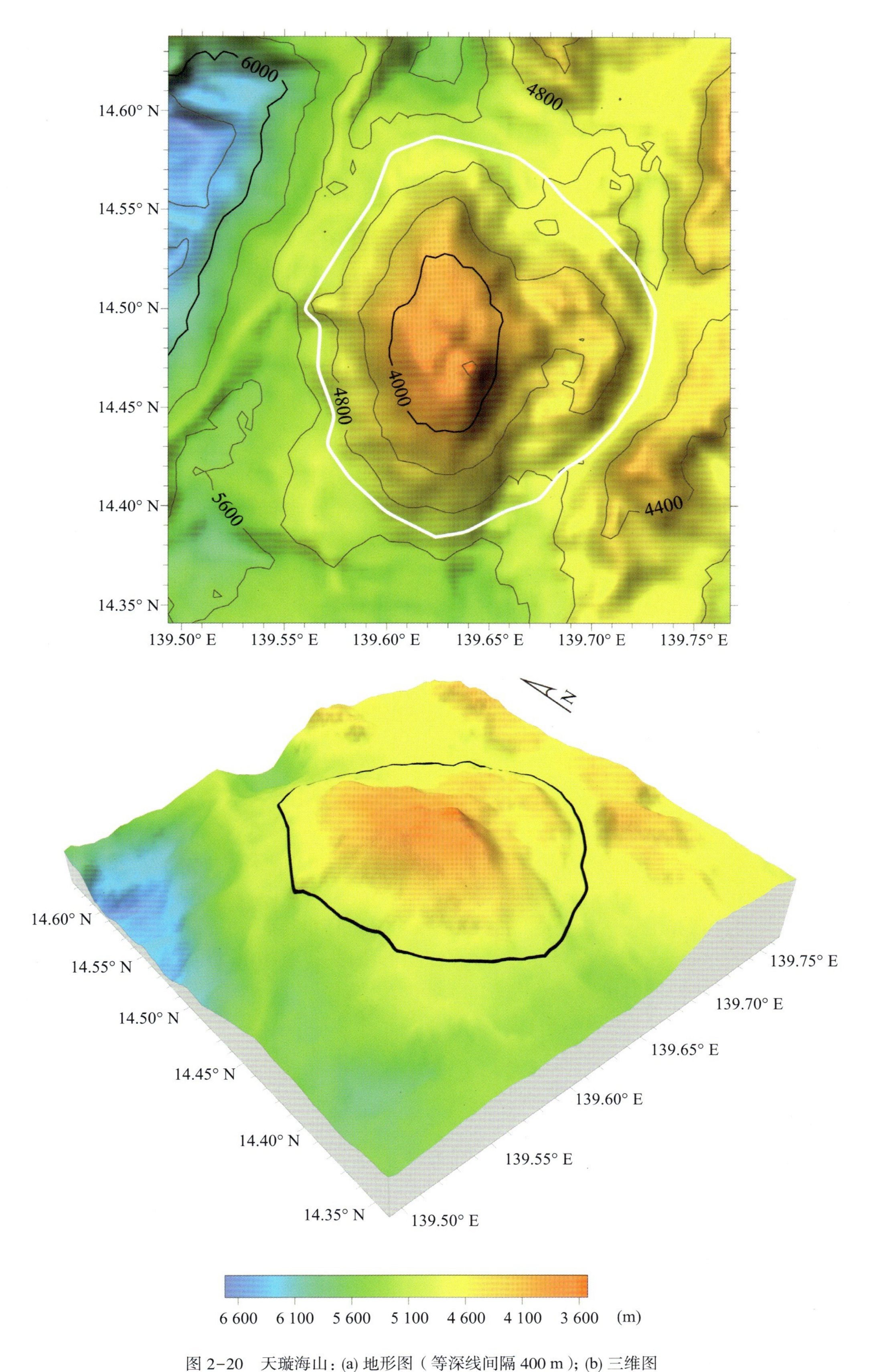

图 2-20 天璇海山：(a) 地形图（等深线间隔 400 m）；(b) 三维图

Fig.2-20 Tianxuan Seamount: (a) Bathymetric map (Contours are in 400 m); (b) 3-D topographic map

2.2.18 天枢海山 Tianshu Seamount

<table>
<tr><td>中文名称
Chinese Name</td><td colspan="3">Tiānshū Hǎishān
天枢海山</td></tr>
<tr><td>英文名称
English Name</td><td colspan="3">Tianshu Seamount</td></tr>
<tr><td>地理区域 / Location</td><td colspan="3">西太平洋菲律宾海盆 The West Philippine Basin</td></tr>
<tr><td rowspan="2">特征点坐标
Coordinate</td><td rowspan="2">13°43.3′N 139°33.5′E</td><td>长度 / Length</td><td>28 km</td></tr>
<tr><td>宽度 / Width</td><td>15 km</td></tr>
<tr><td>水深 / Depth</td><td>2 705~4 552 m</td><td>高差 / Total Relief</td><td>1 847 m</td></tr>
<tr><td>发现情况
Discovery Facts</td><td colspan="3">此海山于 2017 年“向阳红 19”船在执行航次调查时发现。
The seamount was discovered in 2017 during the survey cruise carried out onboard the Chinese R/V Xiang Yang Hong 19.</td></tr>
<tr><td>地形特征
Feature Description</td><td colspan="3">天璇海山位于九州－帕劳海脊以东，帕里西维拉裂谷以南 260 km，位于天璇海山以南 86 km，该海山整体呈 S-N 走向，北窄南宽，西北部陡峭，东南部平缓。
Located to the east of the Kyushu-Palau Ridge, the seamount is 260 km south of the Parece Vela Rift, and 86 km south of Tianxuan Seamount. It has an overall S-N orientation, with a narrow northern end and a wider southern end. The northwestern part is steep, while the southeastern part is gentle.</td></tr>
<tr><td>专名释义
Reason for Choice of Name</td><td colspan="3">名称取自我国传统星官体系“三垣二十八宿”紫薇垣星官中北斗七星的勺头的四颗斗魁星，该海山和其他六个海山依北斗七星分布方位命名，天枢是北斗七星斗魁星中的第一颗星。
The name originates from four bowl stars of the Big Dipper of the Ziwei Enclosure Constellation of the Three Enclosures and Twenty-Eight Constellations in the traditional Chinese constellation system. This feature and other six seamounts are named according to the star distribution of the Big Dipper, and the Tianshu is the first Doukui star of the Big Dipper.</td></tr>
</table>

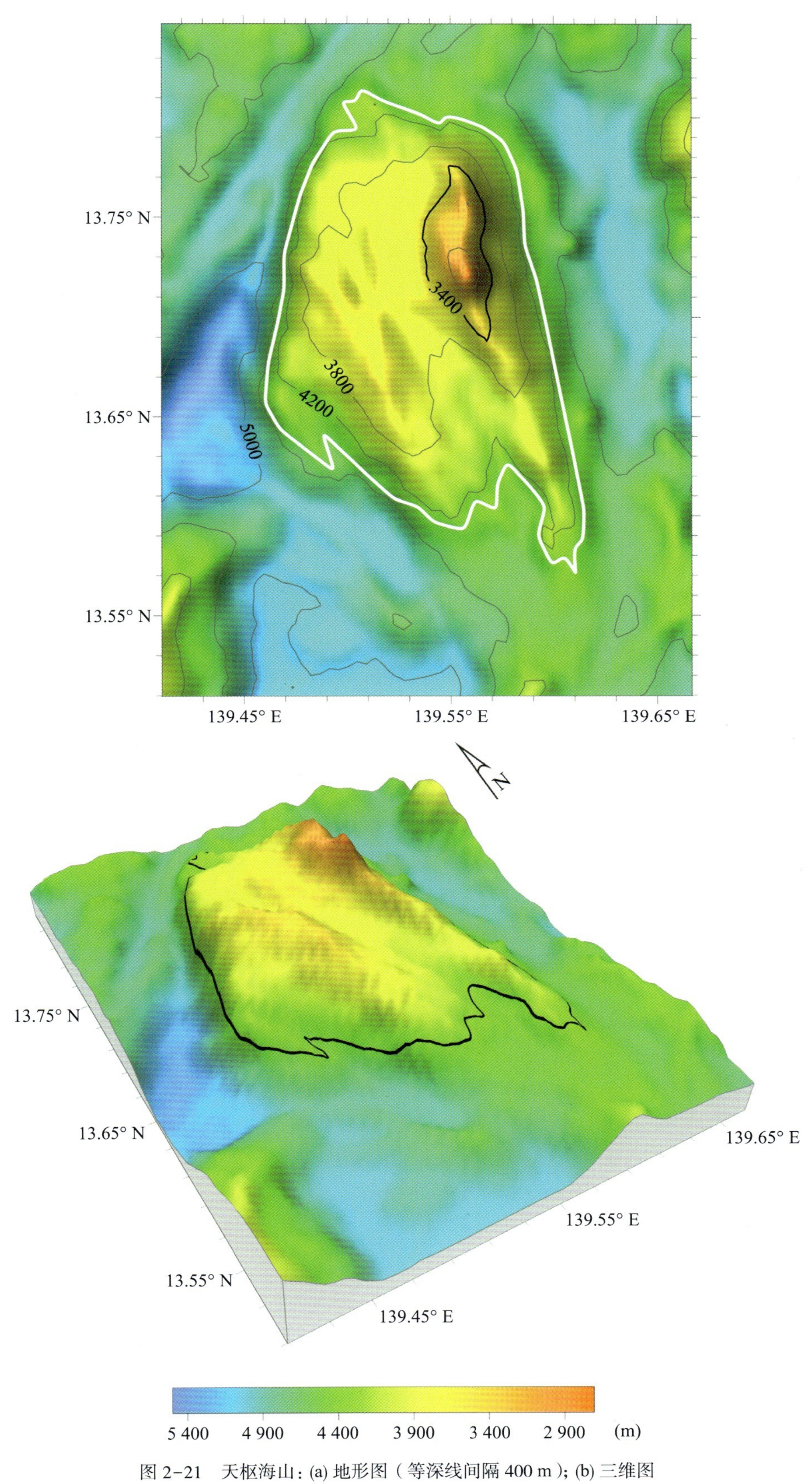

图 2-21 天枢海山：(a) 地形图（等深线间隔 400 m）；(b) 三维图

Fig.2-21 Tianshu Seamount: (a) Bathymetric map (Contours are in 400 m); (b) 3-D topographic map

2.2.19 梗河海山群 Genghe Seamounts

<table>
<tr><td>中文名称
Chinese Name</td><td colspan="3">Gěnghé Hǎishānqún
梗河海山群</td></tr>
<tr><td>英文名称
English Name</td><td colspan="3">Genghe Seamounts</td></tr>
<tr><td>地理区域 / Location</td><td colspan="3">西太平洋菲律宾海盆 The West Philippine Basin</td></tr>
<tr><td rowspan="2">特征点坐标
Coordinate</td><td rowspan="2">14°42.7′N 137°11.8′E
14°42.6′N 137°05.4′E
14°41.5′N 137°07.3′E
14°43.8′N 137°17.4′E</td><td>长度 / Length</td><td>30 km</td></tr>
<tr><td>宽度 / Width</td><td>11 km</td></tr>
<tr><td>水深 / Depth</td><td>3 563~4 765 m</td><td>高差 / Total Relief</td><td>1 202 m</td></tr>
<tr><td>发现情况
Discovery Facts</td><td colspan="3">此海山群于 2014 年“向阳红 14”船在执行航次调查时发现。
The seamounts were discovered in 2014 during the survey cruise carried out onboard the Chinese R/V Xiang Yang Hong 14.</td></tr>
<tr><td>地形特征
Feature Description</td><td colspan="3">梗河海山群位于九州－帕劳海脊以东，帕里西维拉裂谷西南 285 km，整体呈 E-W 走向，由四座海丘组成，彼此之间由明显的鞍部相连，东部海丘形体浑圆，顶部平缓且有明显塌陷，西部三座山体顶部崎岖不平。
Located to the east of the Kyushu-Palau Ridge, the seamounts are 285 km southwest of the Parece Vela Rift. Extending in the E-W direction as a whole, the seamounts consist of four hills, which are connected by obvious saddles. The hill in the east is rounded in shape and has a flat summit with obvious collapses on it, whereas the other 3 hills in the west are rugged and uneven.</td></tr>
<tr><td>专名释义
Reason for
Choice of Name</td><td colspan="3">名称取自我国传统星官体系“三垣二十八宿”东方苍龙第三宿氐宿中的梗河星官，位于现代星座的牧夫座，由三颗水平排列的星星组成，其整体走向与该海山群的走向相似，故命名梗河。
The name originates from the Genghe Asterism in the third constellation of Dongfang Canglong of the Three Enclosures and Twenty-Eight Constellations in the traditional Chinese constellation system. Genghe is part of Bootes according to the modern constellation system. Consisting of 3 stars standing horizontally, the Genghe Asterism extends in the same direction as these features. Therefore, such features are named Genghe Seamounts.</td></tr>
</table>

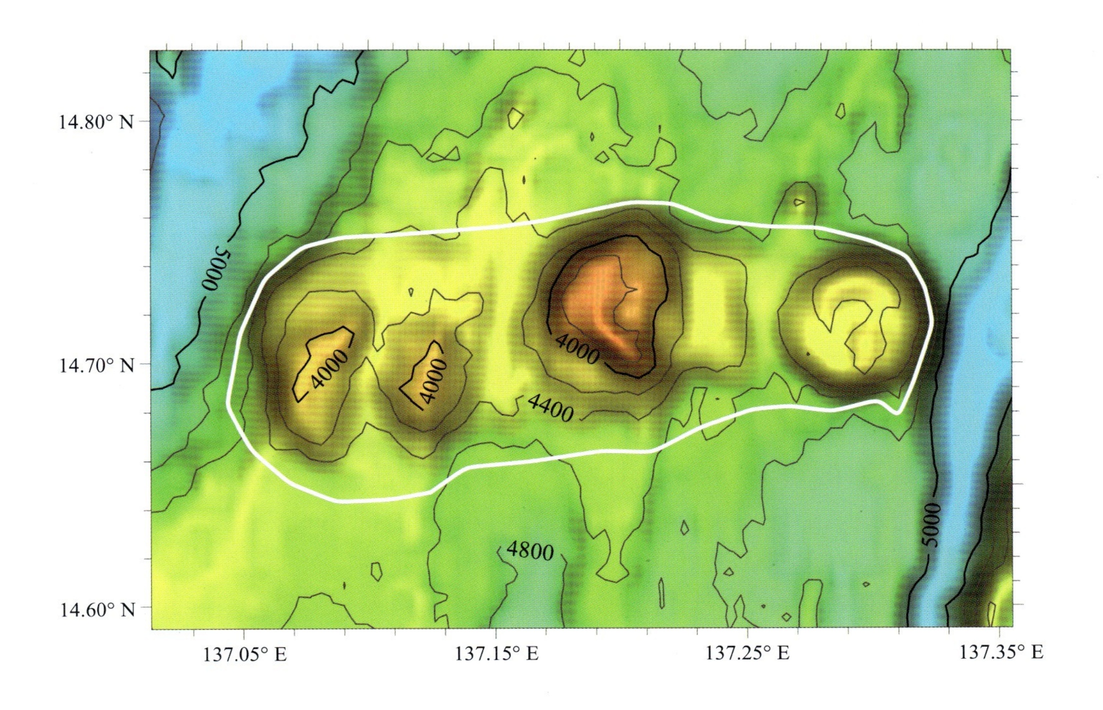

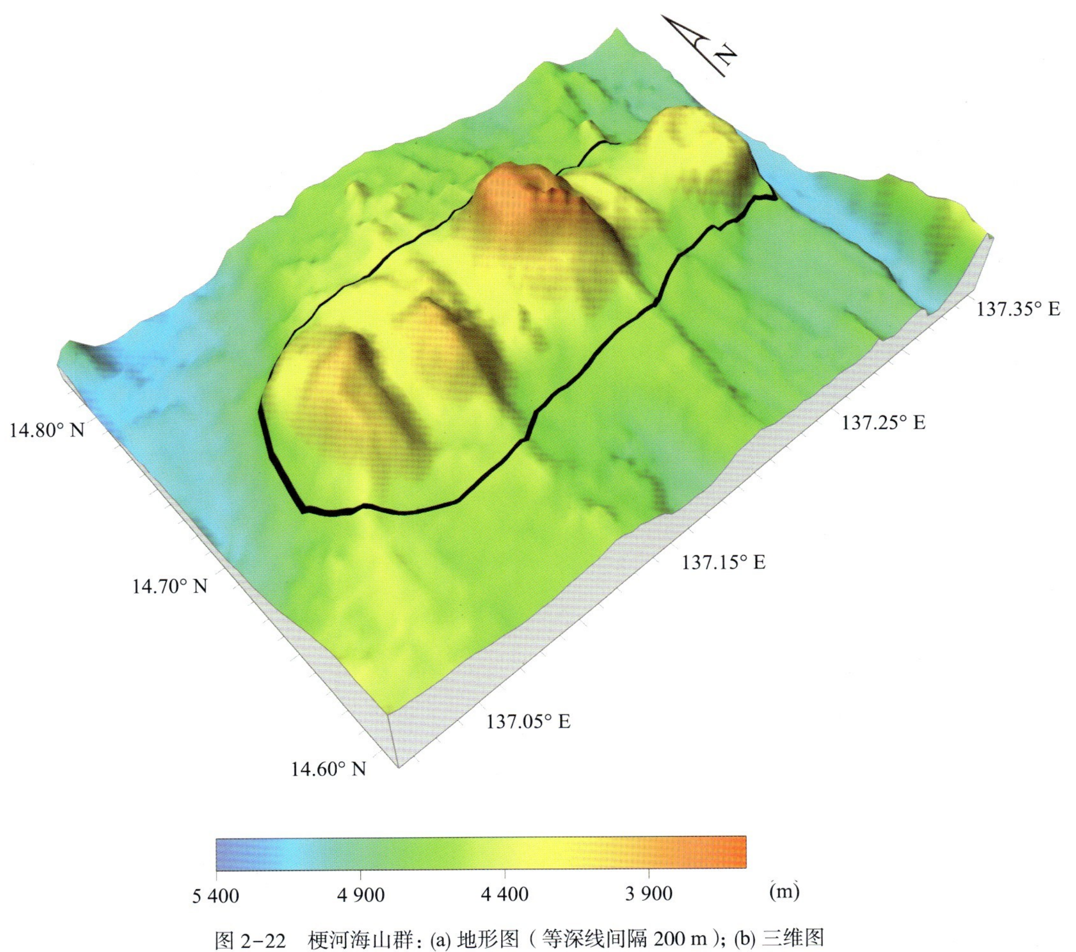

图 2-22 梗河海山群：(a) 地形图（等深线间隔 200 m）；(b) 三维图

Fig.2-22 Genghe Seamounts: (a) Bathymetric map (Contours are in 200 m); (b) 3-D topographic map

2.2.20 发现号海渊 Faxianhao Deep

<table>
<tr><td>中文名称
Chinese Name</td><td colspan="3">Fāxiànhào Hǎiyuān
发现号海渊</td></tr>
<tr><td>英文名称
English Name</td><td colspan="3">Faxianhao Deep</td></tr>
<tr><td>地理区域 / Location</td><td colspan="3">西太平洋菲律宾海盆 The West Philippine Basin</td></tr>
<tr><td rowspan="2">特征点坐标
Coordinate</td><td rowspan="2">13°29.2′N 139°09.7′E</td><td>长度 / Length</td><td>60 km</td></tr>
<tr><td>宽度 / Width</td><td>18 km</td></tr>
<tr><td>水深 / Depth</td><td>5 100~7 026 m</td><td>高差 / Total Relief</td><td>1 926 m</td></tr>
<tr><td>发现情况
Discovery Facts</td><td colspan="3">此海渊于 2017 年“向阳红 19”船在执行航次调查时发现。
The deep was discovered in 2017 during the survey cruise carried out onboard the Chinese R/V Xiang Yang Hong 19.</td></tr>
<tr><td>地形特征
Feature Description</td><td colspan="3">发现号海渊位于帕里西维拉海盆中南部，整体呈 S-N 走向，形体为纺锤状，两端狭窄，中部略宽，中部最宽处约 18 km。该深渊属于帕里西维拉海盆内的裂谷盆地，最深处位于深渊中部，深约 7 026 m。
Located in the central south of the Parece Vela Basin. Extending in the S-N direction as a whole, it takes the shape of a spindle, which is a bit wide in the middle and narrow at both ends, with the widest part being 18 km in the middle. The Deep is a part of the rift basin within the Parece Vela Basin, and the deepest point is located in the middle of it, with a depth of 7,026 m.</td></tr>
<tr><td>专名释义
Reason for
Choice of Name</td><td colspan="3">该海渊以我国 2013 至 2018 年在该海域实施调查的“发现”号深潜器进行命名，用于纪念其在深潜中作出的贡献。
The submersible Faxianhao carried out the undersea investigations in this sea area from 2013 to 2018. Hence the feature is named Faxianhao Deep to commemorate the contribution made by the submersible to the discovering the feature.</td></tr>
</table>

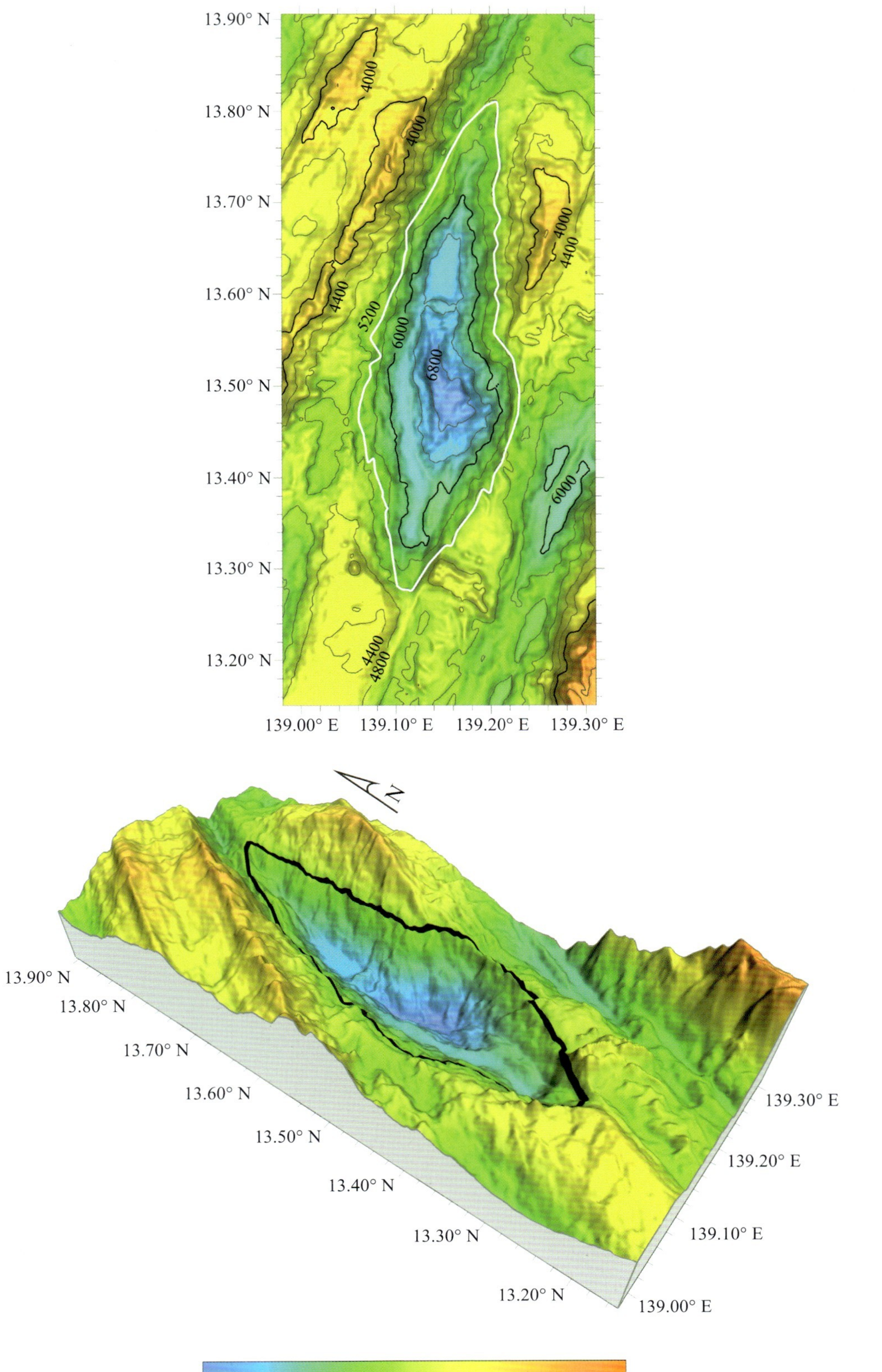

图 2-23 发现号海渊：(a) 地形图（等深线间隔 400 m）；(b) 三维图

Fig.2-23 Faxianhao Deep: (a) Bathymetric map (Contours are in 400 m); (b) 3-D topographic map

3

中北太平洋海底地理实体

Chapter III

Undersea Features in the Central North Pacific Ocean

3.1 地形地貌概况

中北太平洋海域是构造作用的交汇区，周边海域发育大量的海山群、海底高原或海底隆起，地质构造较为复杂，海底地形地貌类型多样。

中北太平洋海域的中部是面积宽广的中太平洋海盆，属于深海洋盆地貌，水深在 5 000~5 500 m 范围内。中太平洋海盆周边为太平洋海山的密集区，其东西两侧分布着一系列北西走向的线性列岛及规模巨大的水下海山链，它们以特定的地理位置及岩石类型构成了较为独特的海山区。紧邻中太平洋海盆北侧为中太平洋海山群，其地理位置为 170°E~165°W，17°N~23°N，位于太平洋海山密集区的中心部位。中太平洋海山群无论是海山排列形态还是走向都与周围海山区有着显著的差异，海山主要呈簇状分布，且呈东西向展布，多为孤立状、双峰状、多顶状平顶海山，是太平洋中海山呈簇状分布密度最高的一个区域。

Section 3.1 Overview of the topography

The Central North Pacific Ocean is a tectonic convergence zone, with a large number of seamounts, plateaus, and seabed uplifts. The seafloor geological structure is complex with a variety of submarine topographic and geomorphologic development types.

The Central Pacific Basin is wide and located in the central part of the Central North Pacific Ocean, which belongs to a deep oceanic basin topography, with depths of 5,000~5,500 m. A large number of seamounts are distributed around the Central Pacific Basin. Especially on both sides of the Central Pacific Basin, there are a series of north-west-trending linear islands and large-scale seamount chains. These seamounts constitute several unique seamount areas with their geological locations and rock types. The Mid-Pacific Seamounts are adjacent to the northern side of the Central Pacific Basin, encompassed between 17°N and 23°N by 170°E and 165°W. The Mid-Pacific Seamounts have significant differences in both their arrangement and trend compared to the surrounding seamounts. These seamounts arrange generally in clusters, and run east to west. The undersea features in this area are mainly guyots with one to two, or multiple peeks.This area is one of the regions in the Pacific Ocean with the highest density of seamount clusters.

3.2 中北太平洋地理实体命名

我国在中北太平洋海域已对 11 个海底地理实体进行了命名，主要以我国古代诗集《诗经》对海底地理实体进行命名。这些海底地理实体主要分布在中太平洋海山群区域。

该海域海底地理实体形态多样，采取太平洋生物群组化命名方法，以象形并结合我国在太平洋海域海洋生物调查发现的物种目、属分类名称对该海域的海底地理实体进行命名。这些新命名的海底地理实体主要分布在中太平洋海盆以南。

新命名的海底地理实体共 15 个，其中平顶山 3 个，为中华鲎平顶山、松骨平顶山、棘骨平顶山；平顶山群 1 个，为鞘群平顶山群；海山 8 个，为茎球海山、星骨海山、简骨海山、瓣棘海山、凤爪海山、飞白枫海山、灯笼鱼海山、梅花参海山；海山群 3 个，为海马海山群、双盘海山群、钳棘海山群。

Section 3.2 Undersea features in the Central North Pacific Ocean

China has already named 11 undersea features in the Central North Pacific Ocean mainly after the ancient Chinese poetry collection *The Book of Songs*. These features are mainly distributed in the Mid-Pacific Seamounts.

The geometric shapes of the undersea features in this area are diverse. A total of 15 newly discovered undersea features in the area are named after groups of Pacific Ocean species, the geometric shapes of the undersea features, the order and genus discovered in surveys. These newly named undersea features are mainly distributed south of the Central Pacific Basin.

These newly discovered undersea features in the Central North Pacific Ocean include 3 guyots named as Zhonghuahou Guyot, Songgu Guyot, and Jigu Guyot; 1 relatively gathering guyots (guyot group) named as Qiaoqun guyots; 8 seamounts named as Jingqiu Seamount, Xinggu Seamount, Jiangu Seamount, Banji Seamount, Fengzhao Seamount, Feibaifeng Seamount, Denglongyu Seamount and Meihuashen Seamount; 3 relatively gathering seamounts (seamount groups) named as Haima Seamounts, Shuangpan Seamounts and Qianji Seamounts.

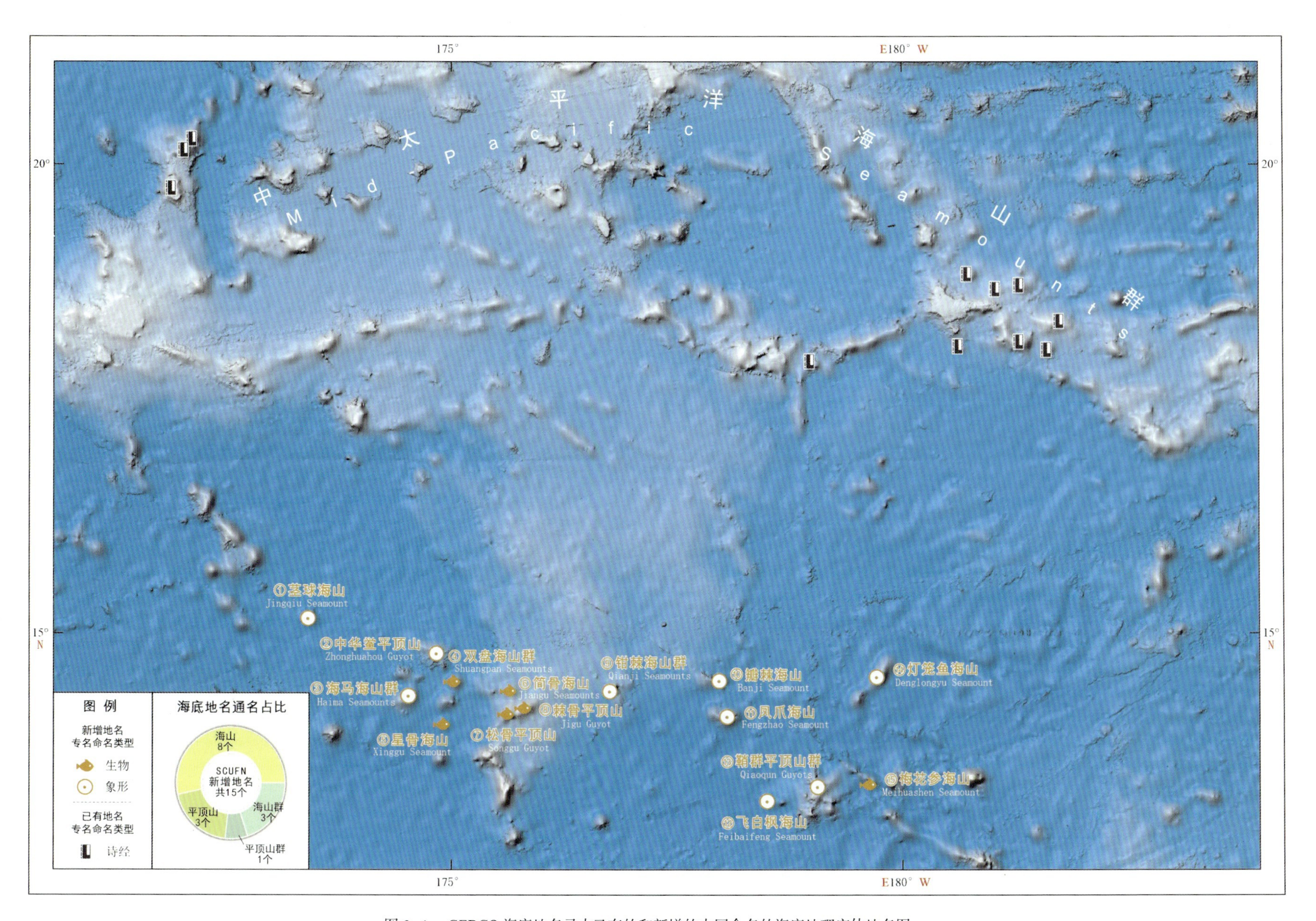

图 3-1 GEBCO 海底地名录中已有的和新增的中国命名的海底地理实体地名图

Fig.3-1 Map of the undersea feature names proposed by China and already included in GEBCO Gazetteer and these newly proposed by China

1	茎球海山	Jingqiu Seamount	15°10.0′N 173°26.1′E
2	中华鲎平顶山	Zhonghuahou Guyot	14°46.4′N 174°51.0′E
3	海马海山群	Haima Seamounts	14°19.5′N 174°31.6′E
4	双盘海山群	Shuangpan Seamounts	14°29.3′N 175°00.3′E
5	星骨海山	Xinggu Seamount	14°01.9′N 174°53.7′E
6	简骨海山	Jiangu Seamount	14°23.2′N 175°37.7′E
7	松骨平顶山	Songgu Guyot	14°08.5′N 175°35.9′E
8	棘骨平顶山	Jigu Guyot	14°12.2′N 175°47.6′E
9	钳棘海山群	Qianji Seamounts	14°23.1′N 176°45.8′E
10	瓣棘海山	Banji Seamount	14°29.1′N 177°58.3′E
11	凤爪海山	Fengzhao Seamount	14°07.0′N 178°03.4′E
12	飞白枫海山	Feibaifeng Seamount	13°13.0′N 178°30.2′E
13	鞘群平顶山群	Qiaoqun Guyots	13°21.4′N 179°03.7′E
14	灯笼鱼海山	Denglongyu Seamount	14°31.7′N 179°43.1′E
15	梅花参海山	Meihuashen Seamount	13°23.4′N 179°37.3′E

3.2.1 茎球海山 Jingqiu Seamount

<table>
<tr><td>中文名称
Chinese Name</td><td colspan="3">Jīngqiú Hǎishān
茎球海山</td></tr>
<tr><td>英文名称
English Name</td><td colspan="3">Jingqiu Seamount</td></tr>
<tr><td>地理区域 / Location</td><td colspan="3">中北太平洋 The Central North Pacific Ocean</td></tr>
<tr><td rowspan="2">特征点坐标
Coordinate</td><td rowspan="2">15°10.0′N 173°26.1′E</td><td>长度 / Length</td><td>19 km</td></tr>
<tr><td>宽度 / Width</td><td>16 km</td></tr>
<tr><td>水深 / Depth</td><td>3 112~5 433 m</td><td>高差 / Total Relief</td><td>2 321 m</td></tr>
<tr><td>发现情况
Discovery Facts</td><td colspan="3">此海山于 2019 年“向阳红 06”船在执行航次调查时发现。
The seamount was discovered in 2019 during the survey cruise carried out onboard the Chinese R/V Xiang Yang Hong 06.</td></tr>
<tr><td>地形特征
Feature Description</td><td colspan="3">茎球海山位于中北太平洋，托马斯平顶山群西南约 255 km，整体呈椭圆形，孤立于深海平原中，东西长 19 km，南北宽 16 km，最高处位于海山中心位置，水深约 3 112 m。
Located in the Central North Pacific Ocean, the seamount is about 255 km southwest of the Thomas Guyots. Being oval-shaped as a whole and isolated on the abyssal plain, the seamount is 19 km long from west to east, and 16 km wide from south to north. The highest point is located in the center of the seamount, with a depth of about 3,112 m.</td></tr>
<tr><td>专名释义
Reason for
Choice of Name</td><td colspan="3">该地区新发现的海底地理实体均以太平洋生物群组化方法命名。该海山形体浑圆，犹如一只生长在海底的茎球海绵。茎球海绵是“Bologominae sp”种属的中文名称，是一种常见于热带太平洋海床的海绵。“蛟龙”号和“发现”号潜水器在太平洋发现了茎球海绵。故以茎球命名。
The newly discovered undersea features in this area are all named after groups of Pacific Ocean species. The undersea feature has rounded shape, which resembles a bulbous sponge growing on the seafloor. Jingqiu is the Chinese name for the species taxa “Bologominae sp”,which is a kind of sessile sponge usually seen on the seabed of the tropical ocean. The submersibles “Jiaolong” and “Faxian” discovered this kind of sponge in the Pacific Ocean. So the feature is named Jingqiu Seamount.</td></tr>
</table>

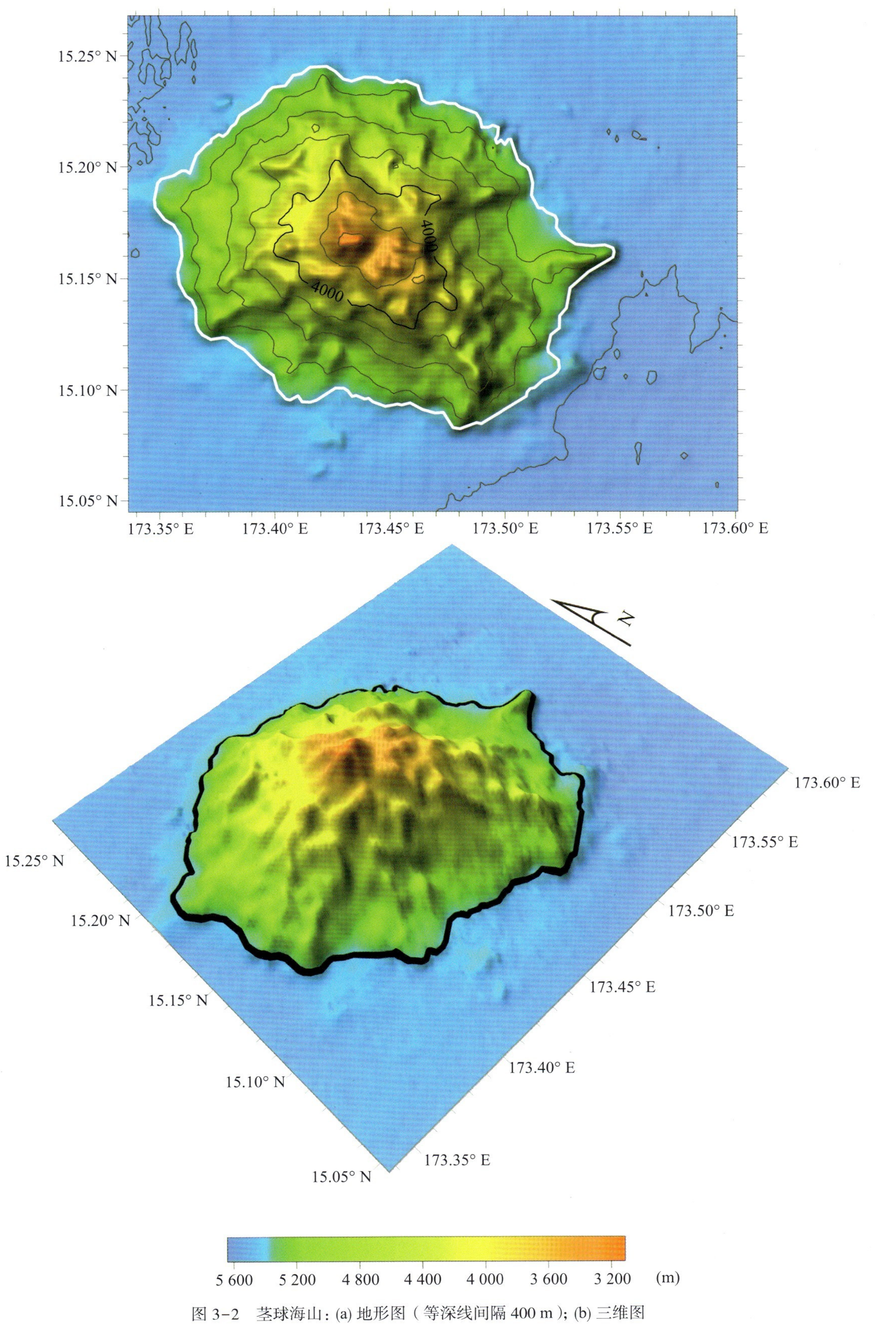

图 3-2 茎球海山：(a) 地形图（等深线间隔 400 m）；(b) 三维图

Fig.3-2 Jingqiu Seamount: (a) Bathymetric map (Contours are in 400 m); (b) 3-D topographic map

3.2.2 中华鲎平顶山 Zhonghuahou Guyot

<table>
<tr><td>中文名称
Chinese Name</td><td colspan="4">Zhōnghuáhòu Píngdǐngshān
中华鲎平顶山</td></tr>
<tr><td>英文名称
English Name</td><td colspan="4">Zhonghuahou Guyot</td></tr>
<tr><td>地理区域 / Location</td><td colspan="4">中北太平洋 The Central North Pacific Ocean</td></tr>
<tr><td rowspan="2">特征点坐标
Coordinate</td><td rowspan="2">14°46.4′N 174°51.0′E</td><td>长度 / Length</td><td>61 km</td></tr>
<tr><td>宽度 / Width</td><td>41 km</td></tr>
<tr><td>水深 / Depth</td><td>2 215~5 189 m</td><td>高差 / Total Relief</td><td>2 974 m</td></tr>
<tr><td>发现情况
Discovery Facts</td><td colspan="4">此平顶山于 2019 年“向阳红 06”船在执行航次调查时发现。
The guyot was discovered in 2019 during the survey cruise carried out onboard the Chinese R/V Xiang Yang Hong 06.</td></tr>
<tr><td>地形特征
Feature Description</td><td colspan="4">中华鲎平顶山孤立于中北太平洋，托马斯平顶山群东南 314 km，呈 NW-SE 走向，东西长 61 km，南北宽 41 km，山体西北向发育脊状延伸。最高峰位于平顶山南侧，峰顶水深 2 215 m。
Being isolated in the Central North Pacific Ocean, the guyot is 314 km southeast of Thomas Guyots. Extending in the SE-NW direction, it is 61 km long from west to east, and 41 km wide from south to north, with the mountain body extending to the northwest like a ridge. The highest peak is located in the south of guyot, with a depth of 2,215 m at the summit.</td></tr>
<tr><td>专名释义
Reason for Choice of Name</td><td colspan="4">俯视该平顶山形似一只在海底匍匐前进的中华鲎。中华鲎，英文名为“Horseshoe Crabs”，拉丁文名为“Tachypleus Tridentatus”，主要分布在中国、印度尼西亚、越南等地，早在 3 亿多年前就生活在地球上，具有“生物活化石”之称，故以中华鲎命名。
The feature looks like a horseshoe crabs crawling on the sea floor from the top view. The Zhonghuahou, also known as Horseshoe Crabs, is called Tachypleus Tridentatus in Latin. The Horseshoe Crabs is mainly distributed in China, Indonesia and Vietnam. It has lived on the earth for more than 300 million years and is known as a “living fossil”. So feature is named Zhonghuahou Guyot.</td></tr>
</table>

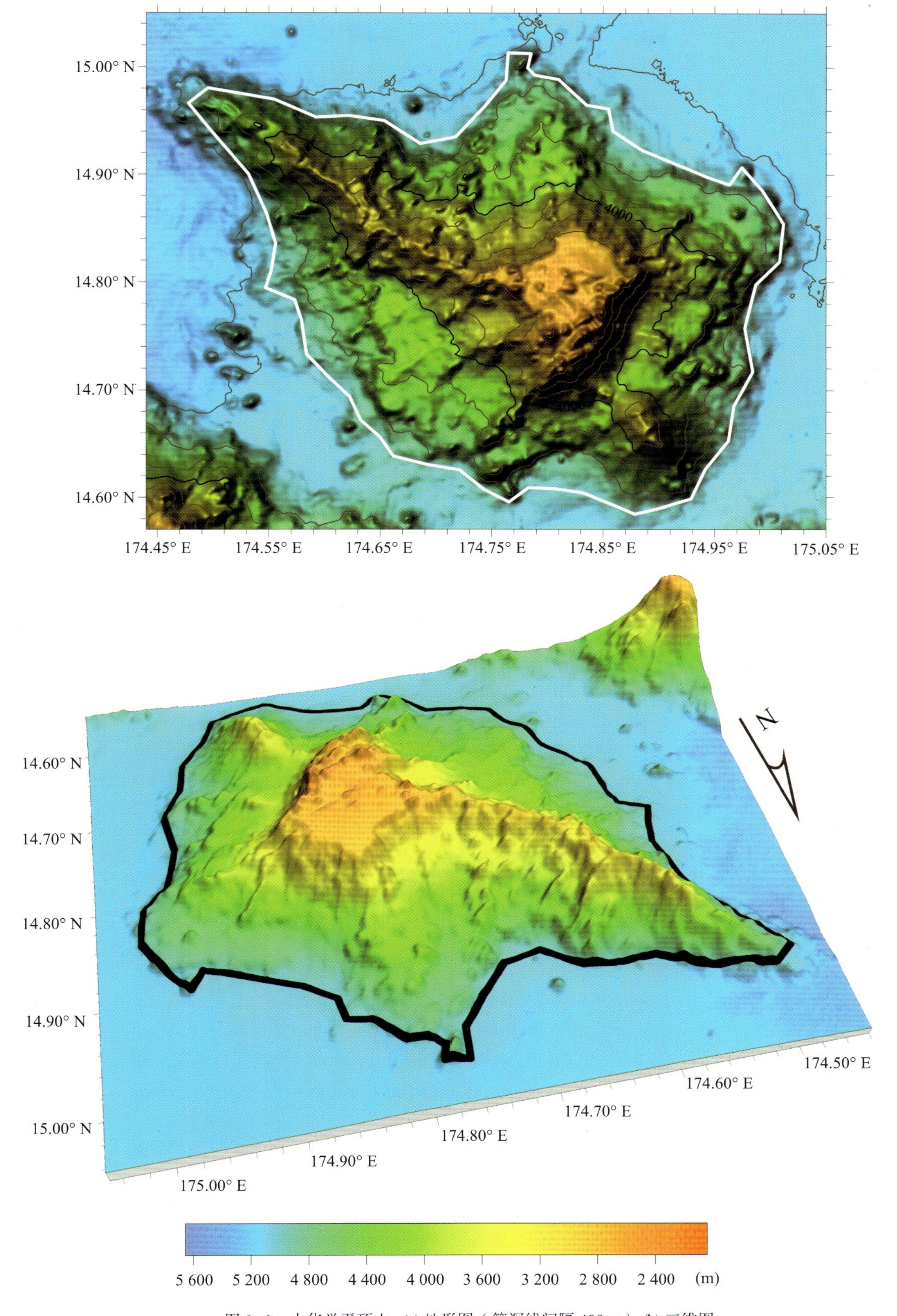

图 3-3 中华鲎平顶山：(a) 地形图（等深线间隔 400 m）；(b) 三维图

Fig.3-3 Zhonghuahou Guyot: (a) Bathymetric map (Contours are in 400 m); (b) 3-D topographic map

3.2.3 海马海山群 Haima Seamounts

中文名称 Chinese Name	Hǎimǎ Hǎishānqún 海马海山群		
英文名称 English Name	Haima Seamounts		
地理区域 / Location	中北太平洋 The Central North Pacific Ocean		
特征点坐标 Coordinate	14°19.5′N 174°31.6′E	**长度 / Length**	54 km
		宽度 / Width	36 km
水深 / Depth	2 064~5 434 m	**高差 / Total Relief**	3 370 m
发现情况 Discovery Facts	此海山群于 2019 年“向阳红 06”船在执行航次调查时发现。 The seamounts were discovered in 2019 during the survey cruise carried out onboard the Chinese R/V *Xiang Yang Hong 06.*		
地形特征 Feature Description	海马海山群位于中北太平洋，托马斯平顶山群东南 353 km，南北长 54 km，东西宽 36 km。山体孤立于深海平原中，发育两个峰顶，一南一北，中间鞍部略微平坦。最高点位于海山群南部，水深为 2 064 m；最深处水深约 5 434 m，位于海山群西缘。 Located in the Central North Pacific Ocean, the seamounts are 353 km southeast of the Thomas Guyots. The seamounts are 54 km long from south to north, and 36 km wide from west to east. Isolated on the abyssal plain, the seamounts have two peaks in the south and the north respectively, with slightly flat saddle in the middle. The highest point is located at the south edge of the seamounts with a depth of 2,064 m, whereas the deepest point is located at the west edge of the seamounts, with a depth of 5,434 m.		
专名释义 Reason for Choice of Name	海马，英文名为“Seahorse”，拉丁文名为“*Hippocampus*”，它们外表独特，有像马一样的头、可卷曲的尾巴和育儿袋。俯视该海山群，北部山峰形似海马之首，南部山峰形似略微卷曲的海马尾部，东北方的凸起使人联想到育儿袋，故以海马命名。 Haima is the Chinese name for seahorse which is also called *Hippocampus* in Latin. They are unique in appearance, with their horse like head, prehensile tail, and brood pouch. The northern seamount looks like the head of a seahorse, the southern seamount looks like the tail of a seahorse, and the northeastern protrusion looks like a brood pouch. So the features are named Haima Seamounts.		

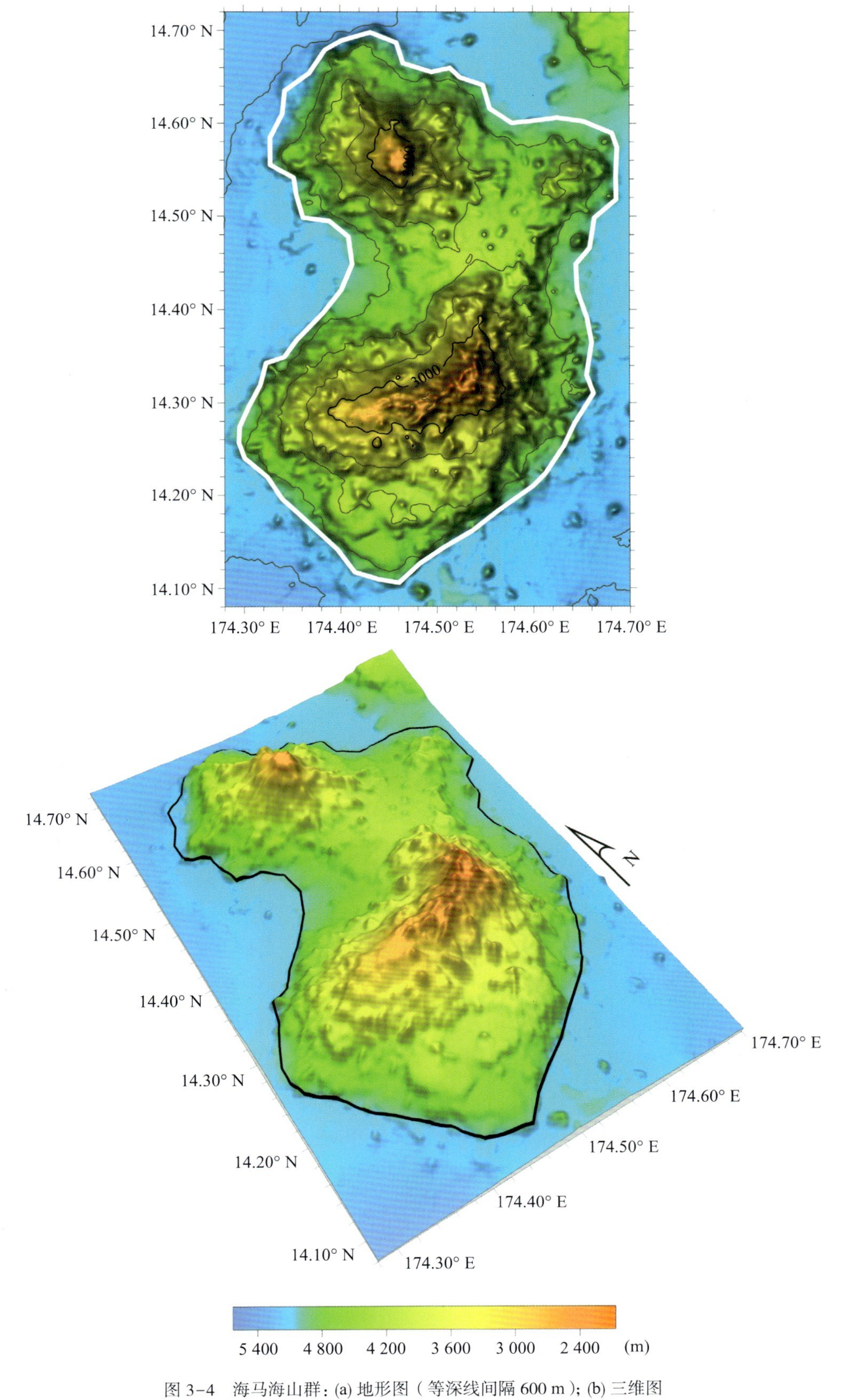

图 3-4 海马海山群：(a) 地形图（等深线间隔 600 m）；(b) 三维图

Fig.3-4 Haima Seamounts: (a) Bathymetric map (Contours are in 600 m); (b) 3-D topographic map

3.2.4 双盘海山群 Shuangpan Seamounts

中文名称 **Chinese Name**	Shuāngpán Hǎishānqún 双盘海山群		
英文名称 **English Name**	Shuangpan Seamounts		
地理区域 / Location	中北太平洋 The Central North Pacific Ocean		
特征点坐标 **Coordinate**	14°29.3′N 175°00.3′E	**长度 / Length**	51 km
		宽度 / Width	25 km
水深 / Depth	2 327~5 279 m	**高差 / Total Relief**	2 952 m
发现情况 **Discovery Facts**	此海山于 2019 年 “向阳红 06” 船在执行航次调查时发现。 The seamounts were discovered in 2019 during the survey cruise carried out onboard the Chinese R/V *Xiang Yang Hong 06*.		
地形特征 **Feature Description**	该海山位于中北太平洋，托马斯平顶山群东南 350 km，呈 NE-SW 走向，南北长 51 km，东西宽 25 km。海山群东部除与一大型海底高地基底相连外，其余方向均为深海平原。其两端各发育独立山峰，外形浑圆，两峰之间为平坦的鞍部，高出周边海底约 500 m。海山群最高处位于北部峰顶，水深为 2 327 m。 Located in the Central North Pacific Ocean, the seamounts are 350 km southeast of the Thomas Guyots. Extending in the SW-NE direction, the seamounts are 51 km long from south to north, and 25 km wide from west to east. Except that the eastern part is connected to the base of a large elevation of the seafloor, all the remaining parts of the seamounts are all abyssal plains. There are rounded independent peaks at both edges of the seamounts, and between the two peaks is a flat saddle, which is about 500 m above the surrounding seafloor. The highest point of the seamounts is located in the northern peak, which is 2,327 m deep.		
专名释义 **Reason for Choice of Name**	该地区新发现的海底地理实体均以太平洋生物群组化方法命名。该海山群发育两个峰顶，在汉语中意为双。双盘，“*Amphidiscosida*”种属的中文名称。“蛟龙”号和“发现”号潜水器在太平洋海域发现了双盘目拟围线海绵。故以双盘命名。 The newly discovered undersea features in this area are all named after groups of the Pacific Ocean species. The undersea features have two mountain tops, which means “shuang” in Chinese. Shuangpan is the Chinese name for the species taxa “*Amphidiscosida*” order. The submersibles “Jiaolong” and “Faxian” discovered this kind of sponge in the Pacific Ocean. So the features are named Shuangpan Seamounts.		

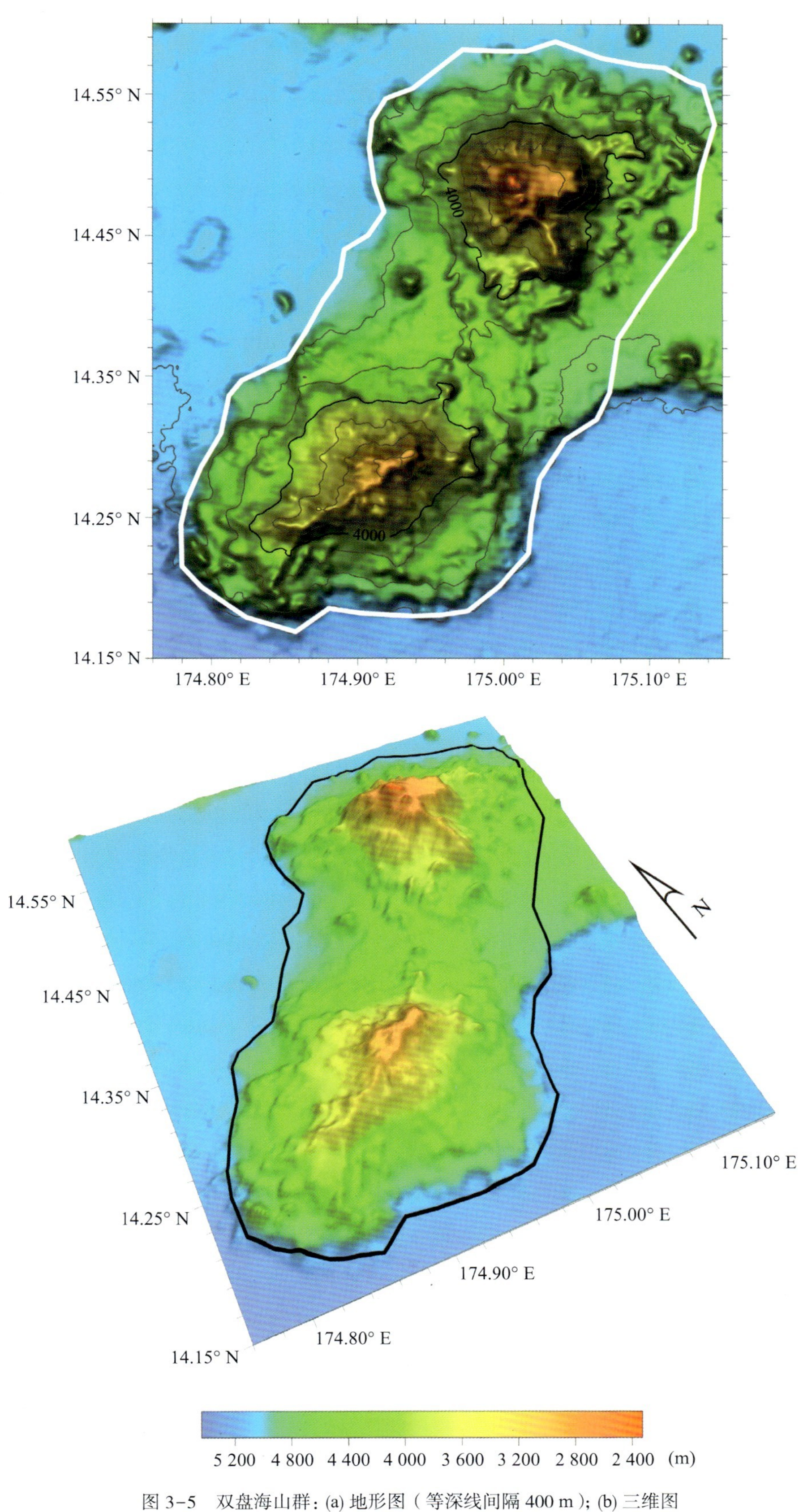

图 3-5 双盘海山群：(a) 地形图（等深线间隔 400 m）；(b) 三维图

Fig.3-5 Shuangpan Seamounts: (a) Bathymetric map (Contours are in 400 m); (b) 3-D topographic map

3.2.5 星骨海山 Xinggu Seamount

<table>
<tr><td>中文名称
Chinese Name</td><td colspan="4">Xīnggǔ Hǎishān
星骨海山</td></tr>
<tr><td>英文名称
English Name</td><td colspan="4">Xinggu Seamount</td></tr>
<tr><td>地理区域 / Location</td><td colspan="4">中北太平洋 The Central North Pacific Ocean</td></tr>
<tr><td rowspan="2">特征点坐标
Coordinate</td><td rowspan="2">14°01.9′N 174°53.7′E</td><td>长度 / Length</td><td colspan="2">15 km</td></tr>
<tr><td>宽度 / Width</td><td colspan="2">11.6 km</td></tr>
<tr><td>水深 / Depth</td><td>3 625~5 431 m</td><td>高差 / Total Relief</td><td colspan="2">1 806 m</td></tr>
<tr><td>发现情况
Discovery Facts</td><td colspan="4">此海山于 2019 年“向阳红 06”船在执行航次调查时发现。
The seamount was discovered in 2019 during the survey cruise carried out onboard the Chinese R/V Xiang Yang Hong 06.</td></tr>
<tr><td>地形特征
Feature Description</td><td colspan="4">星骨海山位于中北太平洋，托马斯平顶山群东南 396 km。该海山整体呈等维展布，孤立于深海平原之上，东西长 15 km，南北宽 11.6 km，最浅处水深约 3 625 m，位于海山中心位置，最大高差约 1 806 m。
Located in the Central North Pacific Ocean, the seamount is 396 km southeast of the Thomas Guyots. Being equidimentional in plan as a whole, it is isolated on the abyssal plain, with a length of 15 km from west to east, and a width of 11.6 km from south to north. The shallowest point, which is at the center of the seamount, is about 3,625 m deep, with a maximum total relief of about 1,806 m.</td></tr>
<tr><td>专名释义
Reason for Choice of Name</td><td colspan="4">该地区新发现的海底地理实体均以太平洋生物群组化方法命名。星骨，“Astrophorina”种属的中文名称。“蛟龙”号和“发现”号潜水器在太平洋发现了星骨亚目类海绵。因此，该海山以星骨命名。
The newly discovered undersea features in this area are all named after groups of the Pacific Ocean species. Xinggu is the Chinese name for the species taxa “Astrophorina” suborder. The submersibles “Jiaolong” and “Faxian” discovered Xinggu suborder sponges in the Pacific Ocean. So the feature is named Xinggu Seamount.</td></tr>
</table>

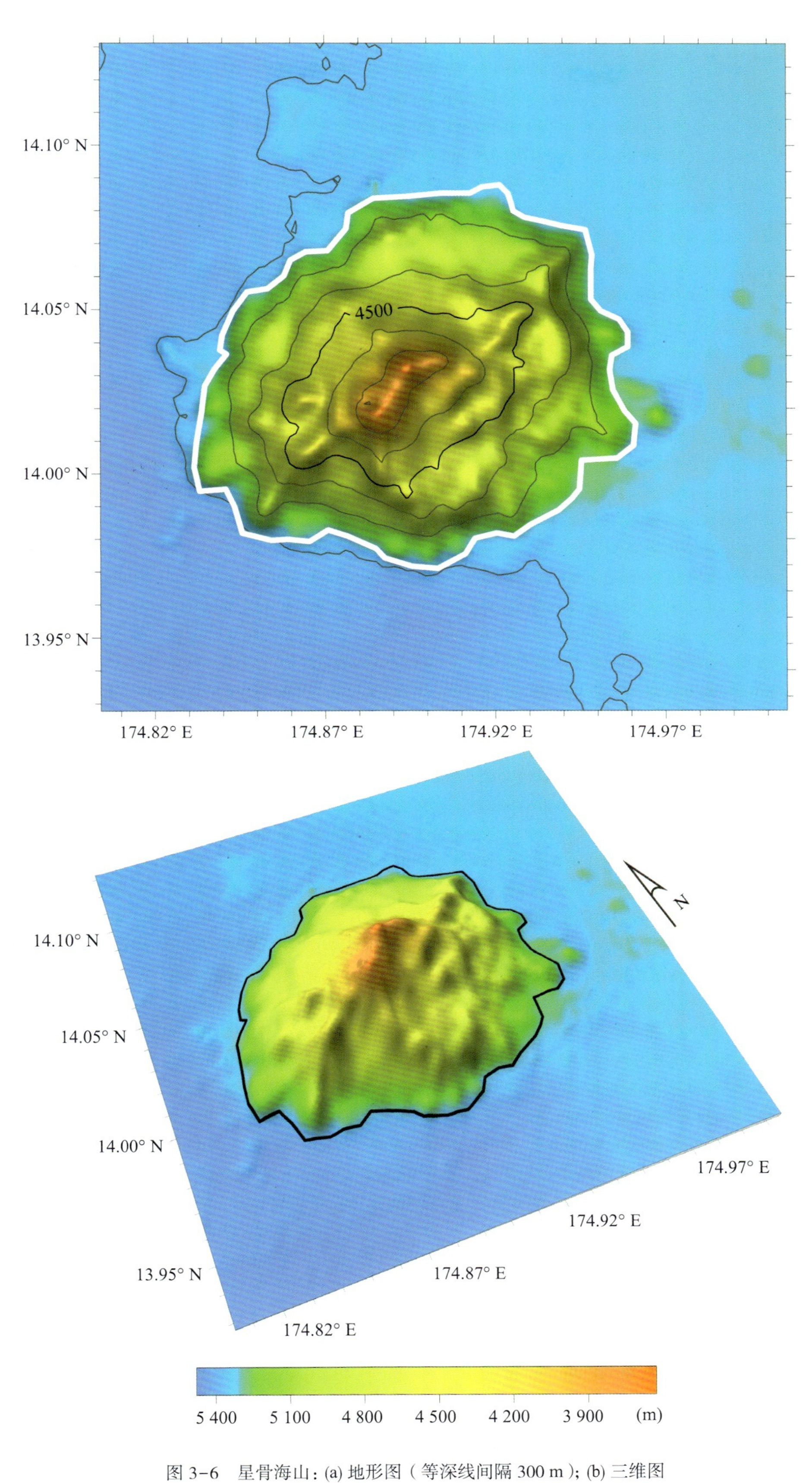

图 3-6 星骨海山：(a) 地形图（等深线间隔 300 m）；(b) 三维图

Fig.3-6 Xinggu Seamount: (a) Bathymetric map (Contours are in 300 m); (b) 3-D topographic map

3.2.6 简骨海山 Jiangu Seamount

<table>
<tr><td>中文名称
Chinese Name</td><td colspan="3">Jiǎngǔ Hǎishān
简骨海山</td></tr>
<tr><td>英文名称
English Name</td><td colspan="3">Jiangu Seamount</td></tr>
<tr><td>地理区域 / Location</td><td colspan="3">中北太平洋 The Central North Pacific Ocean</td></tr>
<tr><td rowspan="2">特征点坐标
Coordinate</td><td rowspan="2">14°23.2′N 175°37.7′E</td><td>长度 / Length</td><td>25 km</td></tr>
<tr><td>宽度 / Width</td><td>14 km</td></tr>
<tr><td>水深 / Depth</td><td>1 569~3 952 m</td><td>高差 / Total Relief</td><td>2 383 m</td></tr>
<tr><td>发现情况
Discovery Facts</td><td colspan="3">此海山于 2019 年“向阳红 06”船在执行航次调查时发现。
The seamount was discovered in 2019 during the survey cruise carried out onboard the Chinese R/V Xiang Yang Hong 06.</td></tr>
<tr><td>地形特征
Feature Description</td><td colspan="3">简骨海山位于中北太平洋，托马斯平顶山群东南 329 km，呈 NE-SW 走向。该海山整体呈等维展布，发育于大型海底高原基底之上，长 25 km，宽 14 km，顶部至四周有辐射状山脊分布。最高处位于海山中心位置，水深约 1 569 m，最大高差可达 2 383 m。该海山除东南向发育高差更大的海山外，其他方向为地形平坦的深海平原。
Located in the Central North Pacific Ocean, the seamount is 329 km southeast of Thomas Guyots, and extends in the NE-SW direction. Being equidimensional in plan as a whole, it stands on the base of a large plateau, and is 25 km long and 14 km wide. There are ridges radiating from the top to the surroundings. The highest point is located in the center of the seamount, with a depth of about 1,569 m and a maximum total relief of 2,383 m. Except the seamount with a greater total relief in the southeast, the remaining parts of Jiangu Seamount are all flat abyssal plains.</td></tr>
<tr><td>专名释义
Reason for Choice of Name</td><td colspan="3">该地区新发现的海底地理实体均以太平洋生物群组化方法命名。简骨，“Haplosclerida”种属的中文名称。“蛟龙”号和“发现”号潜水器在太平洋发现了简骨目海绵。故以简骨命名。
The newly discovered undersea features in this sea area are all named after groups of the Pacific Ocean species. Jiangu is the Chinese name for the species taxa “Haplosclerida” order. The submersibles “Jiaolong” and “Faxian” have discovered the Jiangu sponges in the Pacific Ocean. So the feature is named Jiangu Seamount.</td></tr>
</table>

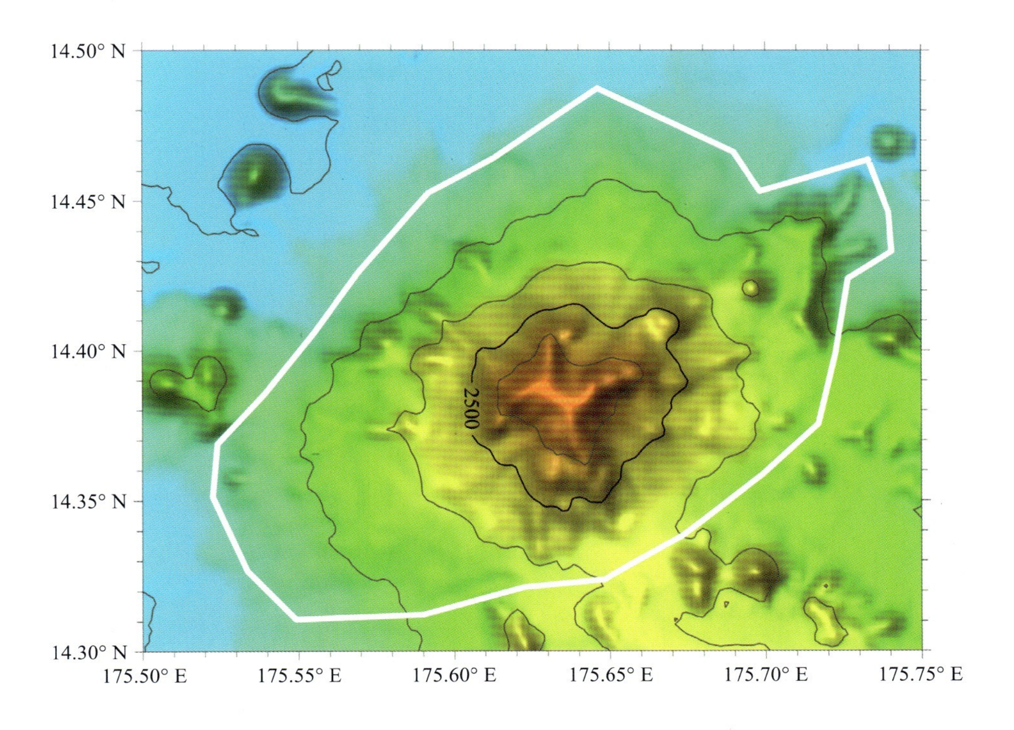

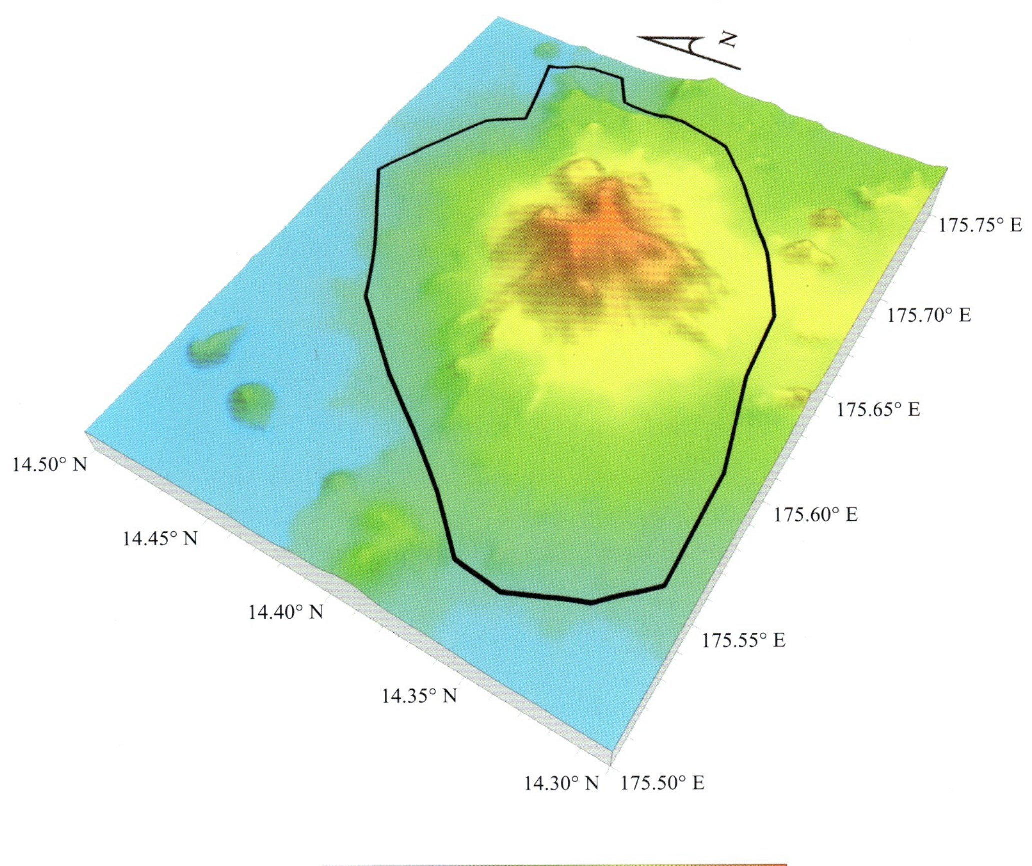

图 3-7 简骨海山：(a) 地形图（等深线间隔 300 m）；(b) 三维图

Fig.3-7 Jiangu Seamount: (a) Bathymetric map (Contours are in 300 m); (b) 3-D topographic map

3.2.7 松骨平顶山 Songgu Guyot

<table>
<tr><td>中文名称
Chinese Name</td><td colspan="4">Sōnggǔ Píngdǐngshān
松骨平顶山</td></tr>
<tr><td>英文名称
English Name</td><td colspan="4">Songgu Guyot</td></tr>
<tr><td>地理区域 / Location</td><td colspan="4">中北太平洋 The Central North Pacific Ocean</td></tr>
<tr><td rowspan="2">特征点坐标
Coordinate</td><td rowspan="2">14°08.5′N 175°35.9′E</td><td>长度 / Length</td><td colspan="2">30 km</td></tr>
<tr><td>宽度 / Width</td><td colspan="2">17 km</td></tr>
<tr><td>水深 / Depth</td><td>1 365~3 746 m</td><td>高差 / Total Relief</td><td colspan="2">2 381 m</td></tr>
<tr><td>发现情况
Discovery Facts</td><td colspan="4">此平顶山于 2019 年“向阳红 06”船在执行航次调查时发现。
The guyot was discovered in 2019 during the survey cruise carried out onboard the Chinese R/V Xiang Yang Hong 06.</td></tr>
<tr><td>地形特征
Feature Description</td><td colspan="4">松骨平顶山位于中北太平洋，托马斯平顶山群东南 415 km，整体呈 NNE-SSW 走向，山体呈宽条形，发育于大型海底高原基底之上，长 30 km，宽 17 km。该平顶山顶部较为平坦，具有典型平顶山的特征。该平顶山最高处水深为 1 365 m，位于顶部东南缘。
Located in the Central North Pacific Ocean, the guyot is 415 km southeast of Thomas Guyots. Extending in the NNE-SSW direction as a whole, it stands on the base of a large plateau in the shape of a wide strip, and is 30 km long and 17 km wide. Being relatively flat at the top, it has the typical features of a guyot. The deepest point of the guyot, which is 1,365 m deep, is located on the southeastern edge of the summit.</td></tr>
<tr><td>专名释义
Reason for
Choice of Name</td><td colspan="4">该地区新发现的海底地理实体均以太平洋生物群组化方法命名。松骨，“Lyssacinosida”目的中文名称。“蛟龙”号和“发现”号潜水器在太平洋发现了松骨目海绵。故以松骨命名。
The newly discovered undersea features in this area are all named after groups of the Pacific Ocean species. Songgu is the Chinese name for the species taxa “Lyssacinosida” order. The submersibles “Jiaolong” and “Faxian” have discovered Songgu sponges in the Pacific Ocean. So the feature is named Songgu Guyot.</td></tr>
</table>

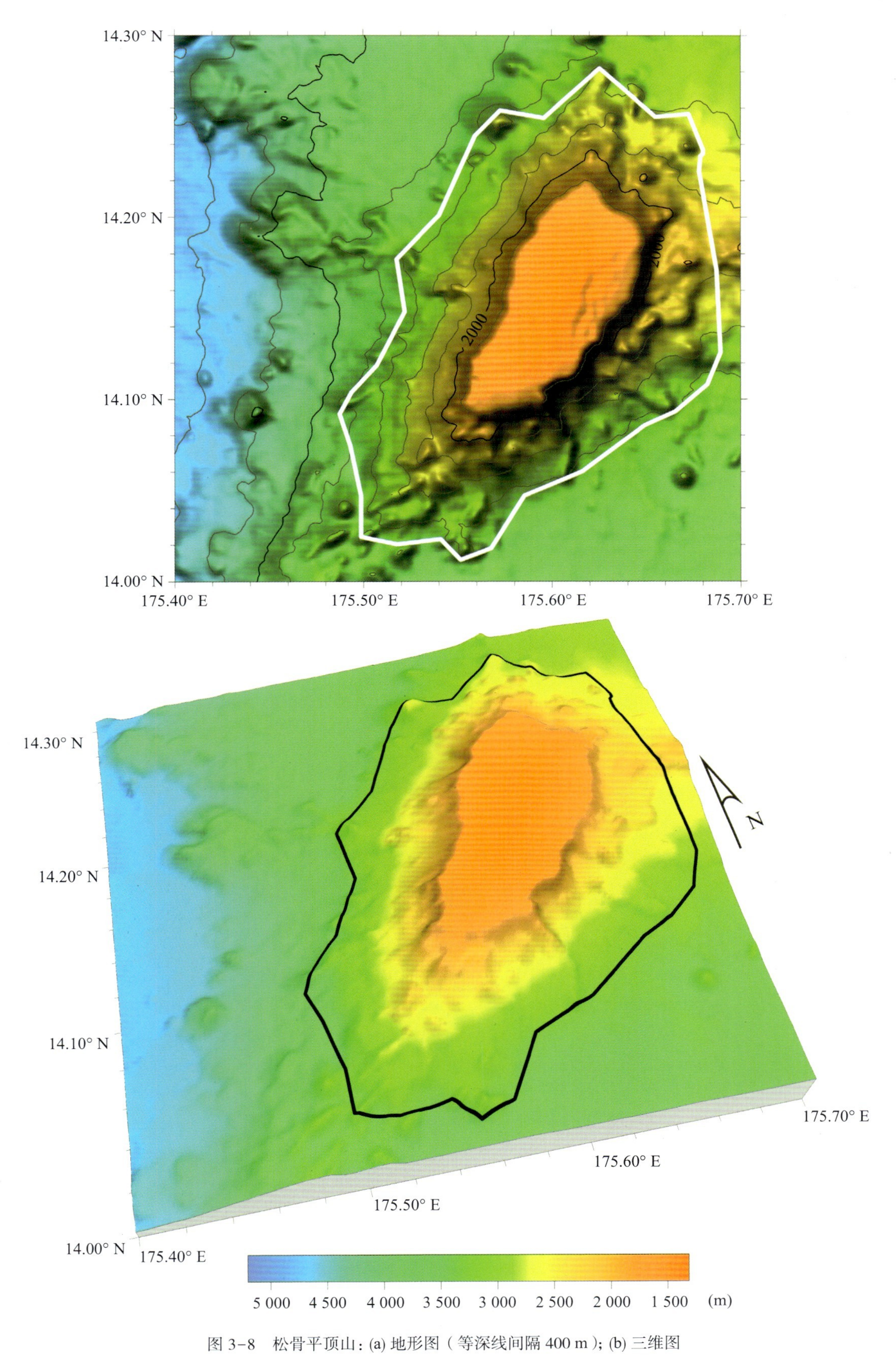

图 3-8 松骨平顶山：(a) 地形图（等深线间隔 400 m）；(b) 三维图

Fig.3-8 Songgu Guyot: (a) Bathymetric map (Contours are in 400 m); (b) 3-D topographic map

3.2.8 棘骨平顶山 Jigu Guyot

中文名称 Chinese Name	Jígǔ Píngdǐngshān 棘骨平顶山		
英文名称 English Name	Jigu Guyot		
地理区域 / Location	中北太平洋 The Central North Pacific Ocean		
特征点坐标 Coordinate	14°12.2′N 175°47.6′E	长度 / Length	38 km
		宽度 / Width	18 km
水深 / Depth	1 314~3 639 m	高差 / Total Relief	2 325 m
发现情况 Discovery Facts	此平顶山于 2019 年“向阳红 06”船在执行航次调查时发现。 The guyot was discovered in 2019 during the survey cruise carried out onboard the Chinese R/V *Xiang Yang Hong 06.*		
地形特征 Feature Description	棘骨平顶山位于中北太平洋，托马斯平顶山群东南 419 km，紧临松骨平顶山。该平顶山呈 NE-SW 走向，发育于大型海底高地基底之上，长 38 km，宽 18 km。该平顶山最高处位于顶部西南边缘，水深为 1 314 m，山坡处广泛发育规模较小的突起。 Located in the Central North Pacific Ocean, the guyot is 419 km southeast of the Thomas Guyots, and close to Songgu Guyot. Standing on the base of a large elevation of the seafloor, it extends in the NE-SW direction, and is 38 km long and 18 km wide. The highest point of the guyot is located at the southwestern edge of the summit, with a depth of 1,314 m. There are small-scale protrusions widely spread on the slope.		
专名释义 Reason for Choice of Name	该地区新发现的海底地理实体均以太平洋生物群组化方法命名。棘骨，“*Acanthascinae*”亚科的中文名称。“蛟龙”号和“发现”号潜水器在太平洋发现了棘骨科海绵。故以棘骨命名。 The newly discovered undersea features in this sea area are all named after groups of the Pacific Ocean species. Jigu is the Chinese name for the species taxa “*Acanthascinae*” subfamily. The submersibles “Jiaolong” and “Faxian” have discovered Jigu sponges in the Pacific Ocean. So the feature is named Jigu Guyot.		

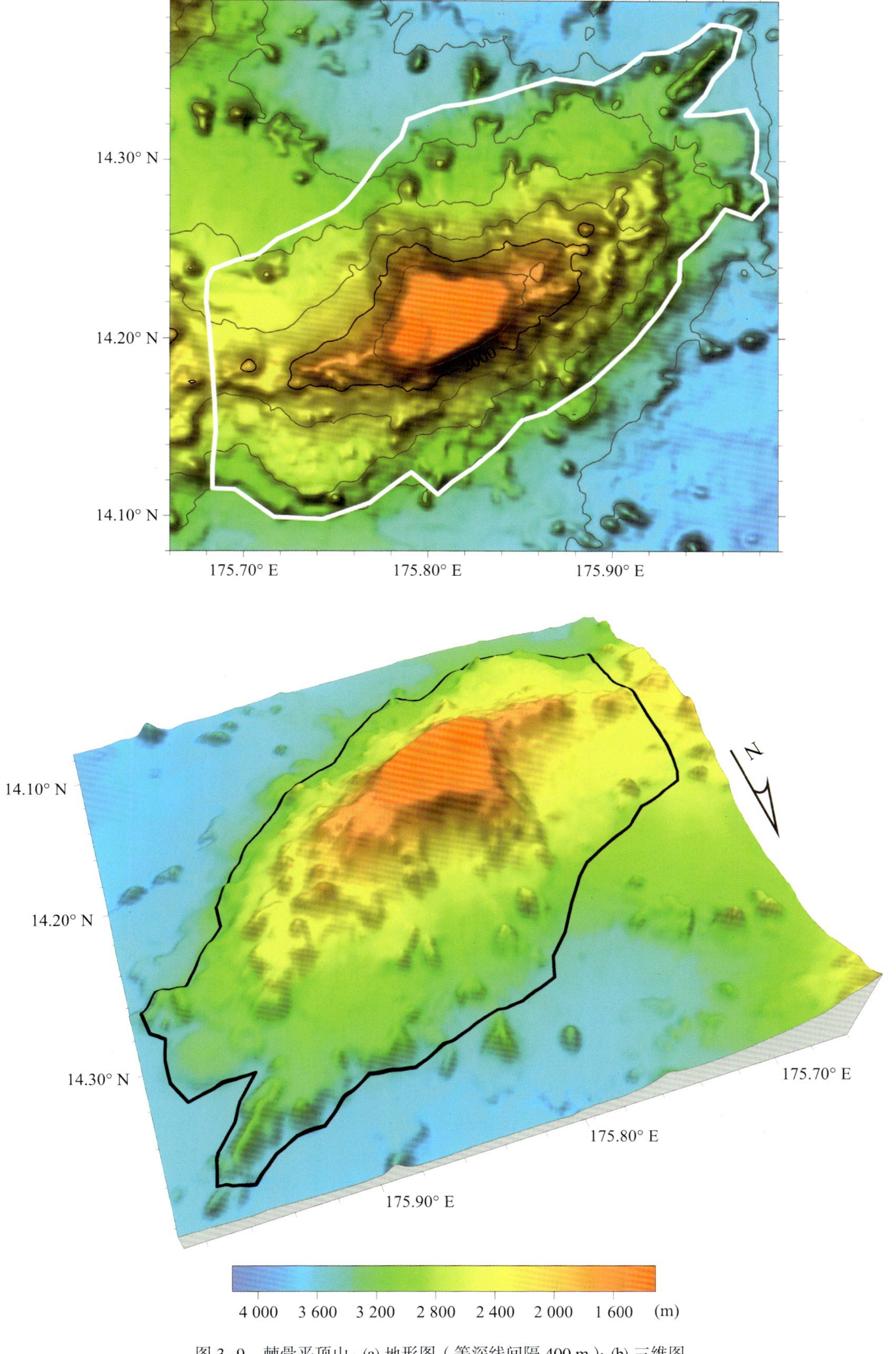

图 3-9　棘骨平顶山：(a) 地形图（等深线间隔 400 m）；(b) 三维图

Fig.3-9　Jigu Guyot: (a) Bathymetric map (Contours are in 400 m); (b) 3-D topographic map

3.2.9 钳棘海山群 Qianji Seamounts

<table>
<tr><td>中文名称
Chinese Name</td><td colspan="3">Qiánjí Hǎishānqún
钳棘海山群</td></tr>
<tr><td>英文名称
English Name</td><td colspan="3">Qianji Seamounts</td></tr>
<tr><td>地理区域 / Location</td><td colspan="3">中北太平洋 The Central North Pacific Ocean</td></tr>
<tr><td rowspan="2">特征点坐标
Coordinate</td><td rowspan="2">14°23.1′N 176°45.8′E</td><td>长度 / Length</td><td>33 km</td></tr>
<tr><td>宽度 / Width</td><td>23 km</td></tr>
<tr><td>水深 / Depth</td><td>1 531~4 532 m</td><td>高差 / Total Relief</td><td>3 001 m</td></tr>
<tr><td>发现情况
Discovery Facts</td><td colspan="3">此海山群于 2019 年“向阳红 14”船在执行航次调查时发现。
The seamounts were discovered in 2019 during the survey cruise carried out onboard the Chinese R/V Xiang Yang Hong 14.</td></tr>
<tr><td>地形特征
Feature Description</td><td colspan="3">钳棘海山群位于中北太平洋，里奥纳尔礁西南 336 km，呈 NW-SE 走向，发育于大型海底高地基底之上，长 33 km，宽 23 km。该海山群由两个山体组成，均呈等维展布，从山顶至四周辐射发育多条小型山脊，其中，北部山体较大，最高处位于中部峰顶，水深为 1 531 m。该海山群坡脚周边广泛发育小型突起。
Located in the Central North Pacific Ocean, the seamounts are 336 km southwest of the Rional Reef, and extend in the NW-SE direction. Standing on the base of a large elevation of the seafloor, the seamounts are 33 km long and 23 km wide. Consisting of two mountain bodies, the seamounts are equidimensional in plan, and have many small ridges radiating from the summit to the surroundings. Among them, the northern mountain body is relatively large, and the highest point is located at the peak in the middle, with a depth of 1,531 m. Small protrusions are widely spread around the foot of the seamounts.</td></tr>
<tr><td>专名释义
Reason for Choice of Name</td><td colspan="3">该地区新发现的海底地理实体均以太平洋生物群组化方法命名。钳棘，“Forcipulatida”种属的中文名称。“蛟龙”号和“发现”号潜水器在太平洋发现了钳棘目海星。该海山群呈辐射状分布多个山脊，俯视形似紧贴于海底的钳棘海星。故以钳棘命名。
The newly discovered undersea features in this sea area are all named after groups of the Pacific Ocean species. Qianji is the Chinese name for the species taxa “Forcipulatida” order. The submersibles “Jiaolong” and “Faxian” have discovered the Qianji starfish in the Pacific Ocean. The undersea features have five radially distributed ridges from the top view, which look like Qianji order starfish clinging to the sea floor. So the features are named as Qianji Seamounts.</td></tr>
</table>

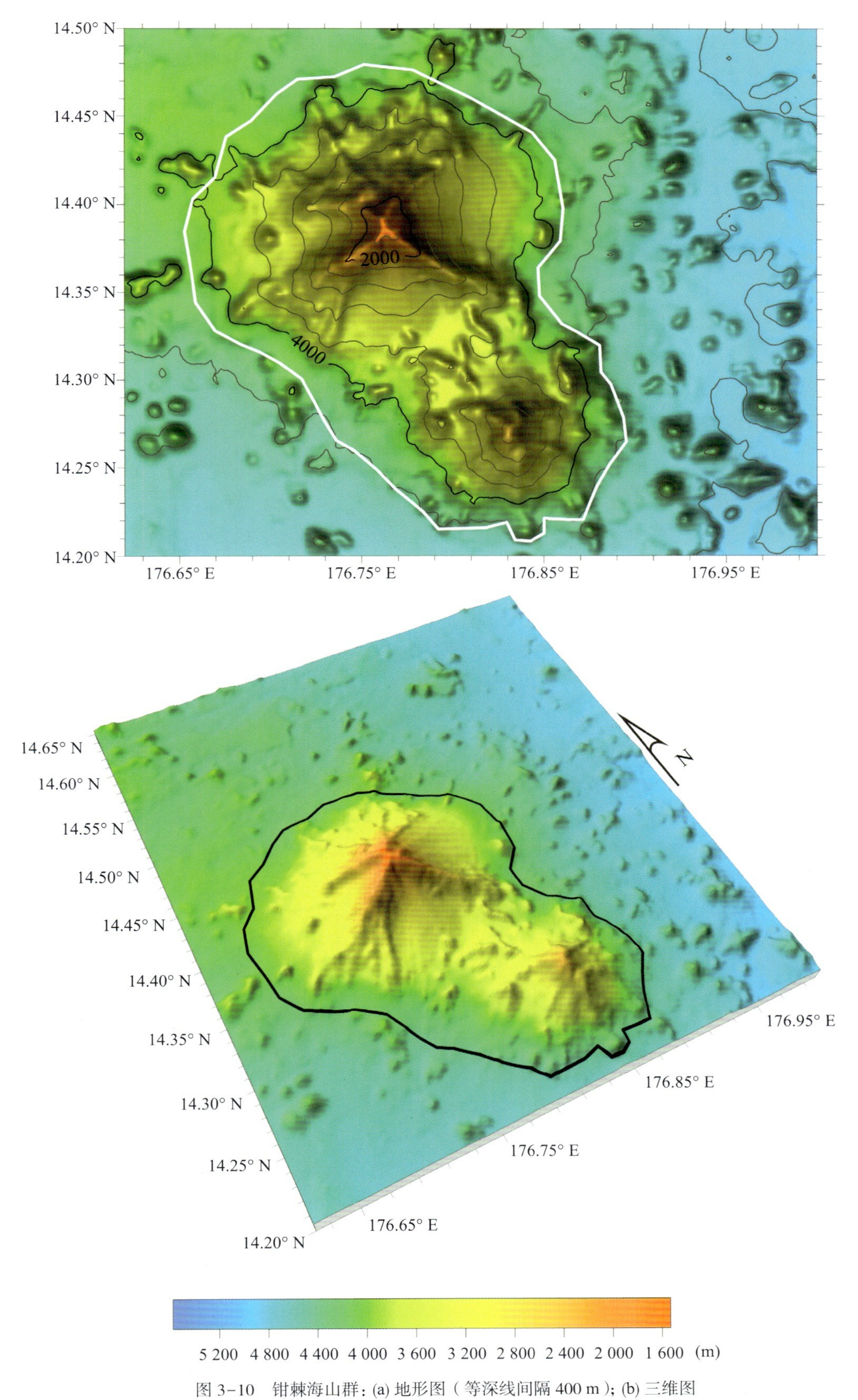

图 3-10 钳棘海山群：(a) 地形图（等深线间隔 400 m）；(b) 三维图

Fig.3-10 Qianji Seamounts: (a) Bathymetric map (Contours are in 400 m); (b) 3-D topographic map

3.2.10 瓣棘海山 Banji Seamount

<table>
<tr><td>中文名称
Chinese Name</td><td colspan="3">Bànjí Hǎishān
瓣棘海山</td></tr>
<tr><td>英文名称
English Name</td><td colspan="3">Banji Seamount</td></tr>
<tr><td>地理区域 / Location</td><td colspan="3">中北太平洋 The Central North Pacific Ocean</td></tr>
<tr><td rowspan="2">特征点坐标
Coordinate</td><td rowspan="2">14°29.1′N 177°58.3′E</td><td>长度 / Length</td><td>20 km</td></tr>
<tr><td>宽度 / Width</td><td>17 km</td></tr>
<tr><td>水深 / Depth</td><td>2 236~5 379 m</td><td>高差 / Total Relief</td><td>3 143 m</td></tr>
<tr><td>发现情况
Discovery Facts</td><td colspan="3">此海山于 2019 年“向阳红 14”船在执行航次调查时发现。
The seamount was discovered in 2019 during the survey cruise carried out onboard the Chinese R/V Xiang Yang Hong 14.</td></tr>
<tr><td>地形特征
Feature Description</td><td colspan="3">瓣棘海山位于中北太平洋，里奥纳尔礁东南 331 km，发育于大型海底高地基底之上，整体呈等维展布，长 20 km，宽 17 km。该海山峰顶位于山体中心，水深为 2 236 m，自峰顶向四周辐射发育六条山脊。该海山坡脚除西北部外，其余方向为平坦的深海平原。
Located in the Central North Pacific Ocean, the seamount is 331 km southeast of the Rional Reef. Standing on the base of a large elevation of the seafloor, it is equidimensional in plan as a whole, with a length of 20 km and a width of 17 km. The peak of the seamount is in the middle of the mountain body, with a depth of 2,236 m. There are 6 ridges radiating from the peak to the surroundings. Except the northwestern part, the remaining parts of the slope foot of the seamount are all flat abyssal plains.</td></tr>
<tr><td>专名释义
Reason for Choice of Name</td><td colspan="3">该地区新发现的海底地理实体均以太平洋生物群组化方法命名。瓣棘，“Valvatida”目的中文名称。“蛟龙”号和“发现”号潜水器在太平洋海域发现了瓣棘目海星。该海山呈辐射状分布多个山脊，俯视形似一只在深海休憩的瓣棘目海星。故以瓣棘命名。
The newly discovered undersea features in this sea area are all named after groups of the Pacific Ocean species. Banji is the Chinese name for the species taxa “Valvatida” order. The submersibles “Jiaolong” and “Faxian” have discovered the Banji starfish in the Pacific Ocean. The undersea feature has radially distributed ridges from the top view, which looks like Banji order starfish clinging to the sea floor. So the undersea feature is named Banji Seamount.</td></tr>
</table>

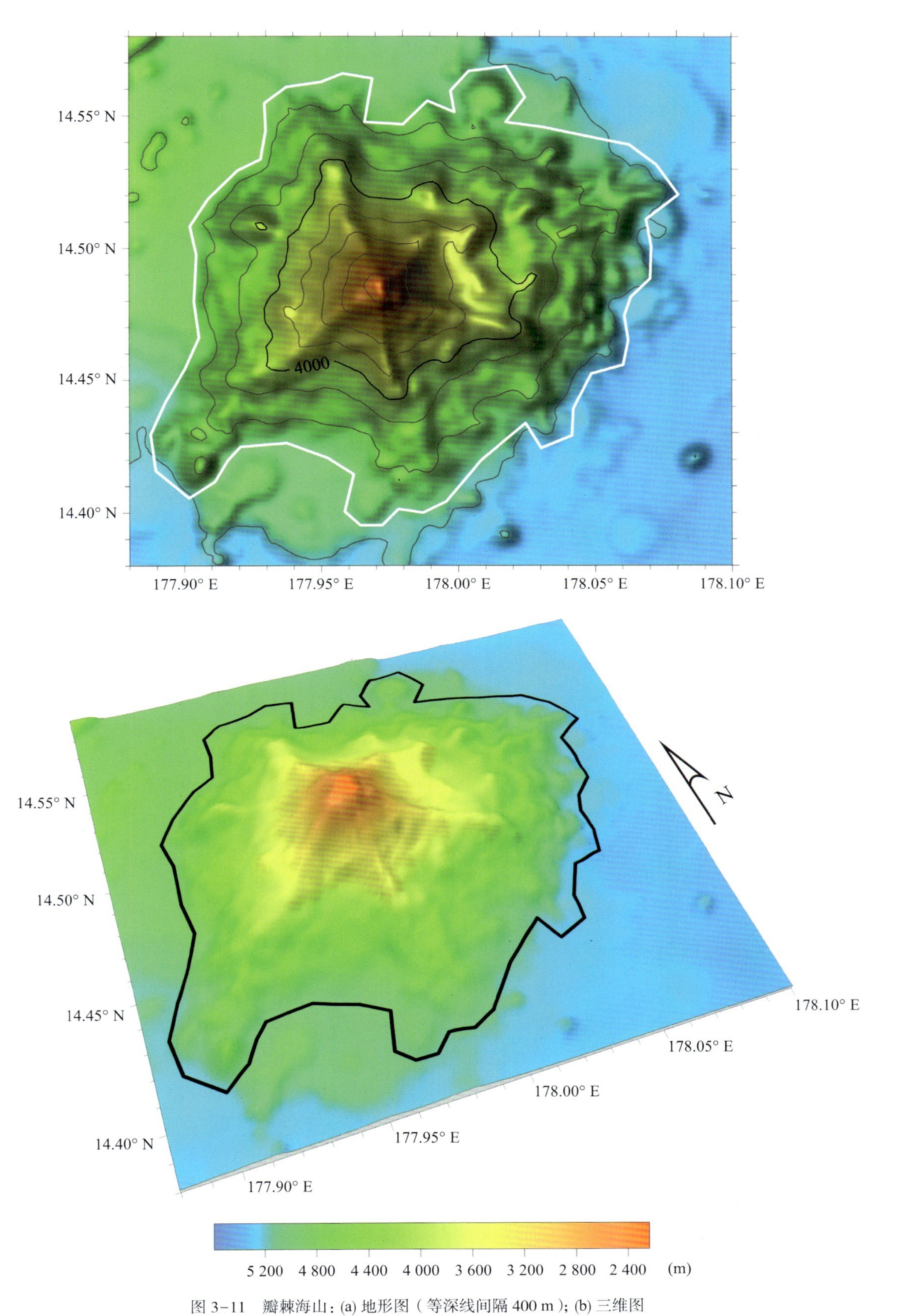

图 3-11 瓣棘海山：(a) 地形图（等深线间隔 400 m）；(b) 三维图

Fig.3-11 Banji Seamount: (a) Bathymetric map (Contours are in 400 m); (b) 3-D topographic map

3.2.11 凤爪海山 Fengzhao Seamount

<table>
<tr><td>中文名称
Chinese Name</td><td colspan="3">Fèngzhǎo Hǎishān
凤爪海山</td></tr>
<tr><td>英文名称
English Name</td><td colspan="3">Fengzhao Seamount</td></tr>
<tr><td>地理区域 / Location</td><td colspan="3">中北太平洋 The Central North Pacific Ocean</td></tr>
<tr><td rowspan="2">特征点坐标
Coordinate</td><td rowspan="2">14°07.0′N 178°03.4′E</td><td>长度 / Length</td><td>45 km</td></tr>
<tr><td>宽度 / Width</td><td>36 km</td></tr>
<tr><td>水深 / Depth</td><td>1 662~5 592 m</td><td>高差 / Total Relief</td><td>3 930 m</td></tr>
<tr><td>发现情况
Discovery Facts</td><td colspan="3">此海山于 2019 年“向阳红 14”船在执行航次调查时发现。
The seamount was discovered in 2019 during the survey cruise onboard the Chinese R/V Xiang Yang Hong 14.</td></tr>
<tr><td>地形特征
Feature Description</td><td colspan="3">凤爪海山位于中北太平洋，托马斯平顶山群东南 594 km，呈 NW-SE 走向，孤立于深海平原之上，东西长 45 km，南北宽 36 km。该海山由两个山体组成，均呈等维展布，从山顶至四周辐射发育多条小型山脊。该海山最高处位于东南部较大的山峰中心，水深为 1 662 m。该海山周围是较为平坦的深海平原，水深超过 5 500 m。
Located in the Central North Pacific Ocean, the seamount is 594 km southeast of the Thomas Guyots, and extends in the NW-SE direction. Being isolated on the abyssal plain, it is 45 km long from west to east, and 36 km wide from south to north. Consisting of two mountain bodies that are equidimensional in plan, the seamount has many small ridges radiating from the peak to the surroundings. The highest point of the seamount is located in the center of a large peak in the southeast, with a depth of 1,662 m. Around the seamount are relatively flat abyssal plains which are more than 5,500 m deep.</td></tr>
<tr><td>专名释义
Reason for Choice of Name</td><td colspan="3">该地区新发现的海底地理实体均以太平洋生物群组化方法命名。鸡爪，“Henricia”种属的中文名称。“蛟龙”号和“发现”号潜水器在太平洋海域发现了鸡爪属海星。该海山从山顶至坡脚呈辐射状分布多个山脊，俯视形似一只紧贴于海底的鸡爪属海星，鸡爪又名凤爪。故以凤爪命名。
The newly discovered undersea features in this sea area are all named after groups of the Pacific Ocean species. Jizhao is the Chinese name for the species taxa “Henricia” genus. The submersibles “Jiaolong” and “Faxian” have discovered Jizhao starfish in the Pacific Ocean. The undersea feature has many ridges distributed radially from the top of the mountain to the foot of the slope, which looks like Jizhao genus starfish clinging to the sea floor from the top view. Jizhao is also called Fengzhao in Chinese. So the feature is named Fengzhao Seamount.</td></tr>
</table>

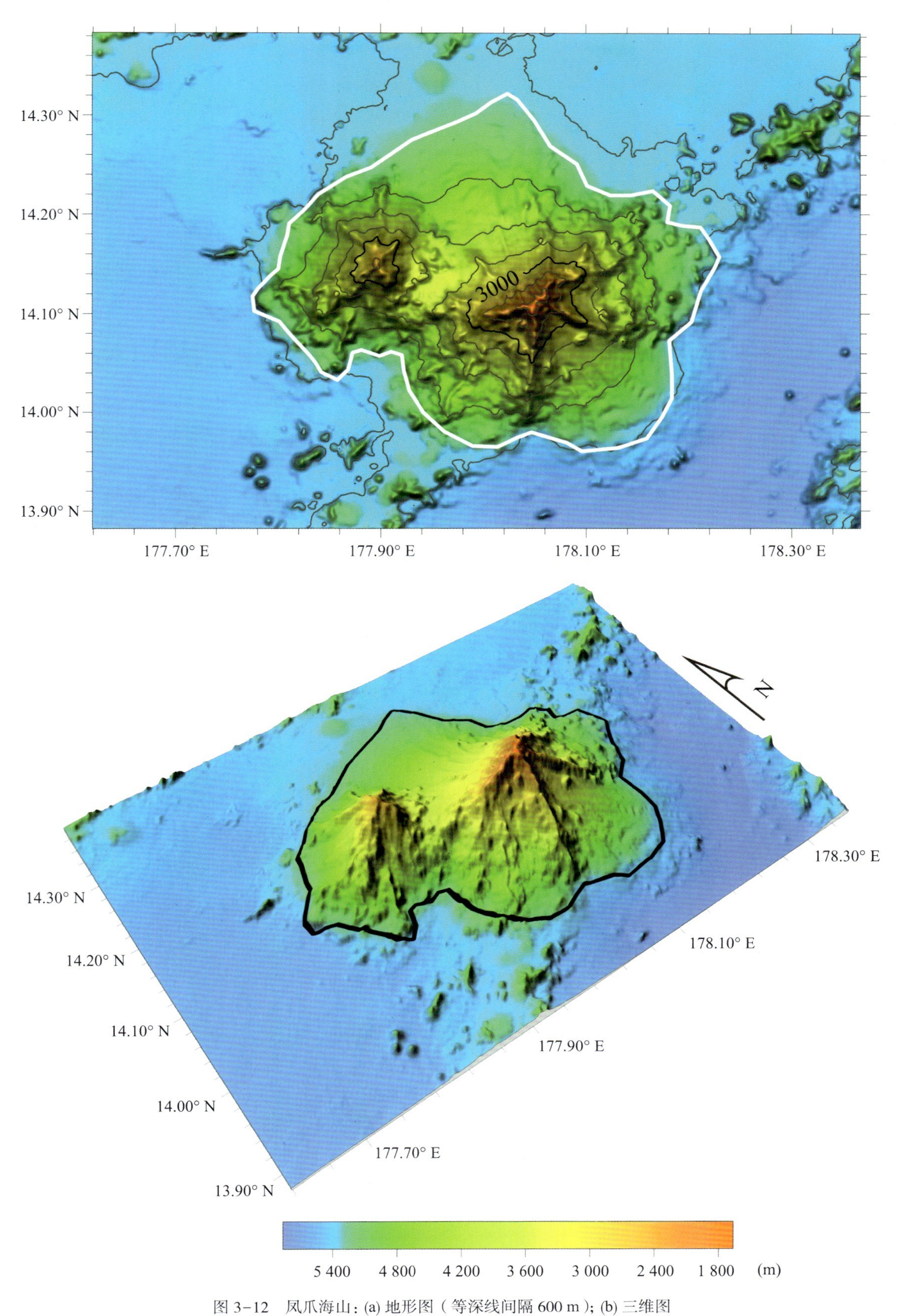

图 3-12 凤爪海山：(a) 地形图（等深线间隔 600 m）；(b) 三维图

Fig.3-12 Fengzhao Seamount: (a) Bathymetric map (Contours are in 600 m); (b) 3-D topographic map

3.2.12 飞白枫海山 Feibaifeng Seamount

<table>
<tr><td>中文名称
Chinese Name</td><td colspan="3">Fēibáifēng Hǎishān
飞白枫海山</td></tr>
<tr><td>英文名称
English Name</td><td colspan="3">Feibaifeng Seamount</td></tr>
<tr><td>地理区域 / Location</td><td colspan="3">中北太平洋 The Central North Pacific Ocean</td></tr>
<tr><td rowspan="2">特征点坐标
Coordinate</td><td rowspan="2">13°13.0′N 178°30.2′E</td><td>长度 / Length</td><td>15 km</td></tr>
<tr><td>宽度 / Width</td><td>13 km</td></tr>
<tr><td>水深 / Depth</td><td>3 887~5 885 m</td><td>高差 / Total Relief</td><td>1 998 m</td></tr>
<tr><td>发现情况
Discovery Facts</td><td colspan="3">此海山于 2019 年“向阳红 03”船在执行航次调查时发现。
The seamount was discovered in 2019 during the survey cruise carried out onboard the Chinese R/V Xiang Yang Hong 03.</td></tr>
<tr><td>地形特征
Feature Description</td><td colspan="3">飞白枫海山位于中北太平洋，勒内礁西南 409 km。该海山形状规则，呈等边三角形，长约 15 km，宽 13 km，最高处位于山体中心，水深约 3 887 m，从山顶至坡脚呈辐射状发育五条山脊。该海山孤立于深海平原中，周围地形较为平坦，水深超过 5 800 m。
Located in the Central North Pacific Ocean, the seamount is 409 km southwest of the Rene Reef. In the shape of a regular equilateral triangle, it is about 15 km long and 13 km wide. The highest point is located in the center of the mountain body, with a depth of about 3,887 m. Five ridges are radiating from the top of the seamount to the foot of its slope. The seamount is isolated on the abyssal plain, and it is relatively flat in its surroundings, with a depth exceeding 5,800 m.</td></tr>
<tr><td>专名释义
Reason for Choice of Name</td><td colspan="3">该地区新发现的海底地理实体均以太平洋生物群组化方法命名。飞白枫，“Pseudarchasteridae”科的中文名称。“蛟龙”号和“发现”号潜水器在太平洋发现了飞白枫科海星，该海山呈辐射状发育五个山脊，俯视形似一只紧贴于海底的飞白枫海星。故以飞白枫命名。
The newly discovered undersea features in this sea area are all named after groups of Pacific Ocean species. Feibaifeng is the Chinese name for the species taxa “Pseudarchasteridae” family. The submersibles “Jiaolong” and “Faxian” have discovered Feibaifeng starfish in the Pacific Ocean. The undersea feature has five radially distributed ridges from the top view, which looks like Feibaifeng family starfish clinging to the sea floor. So the feature is named Feibaifeng Seamount.</td></tr>
</table>

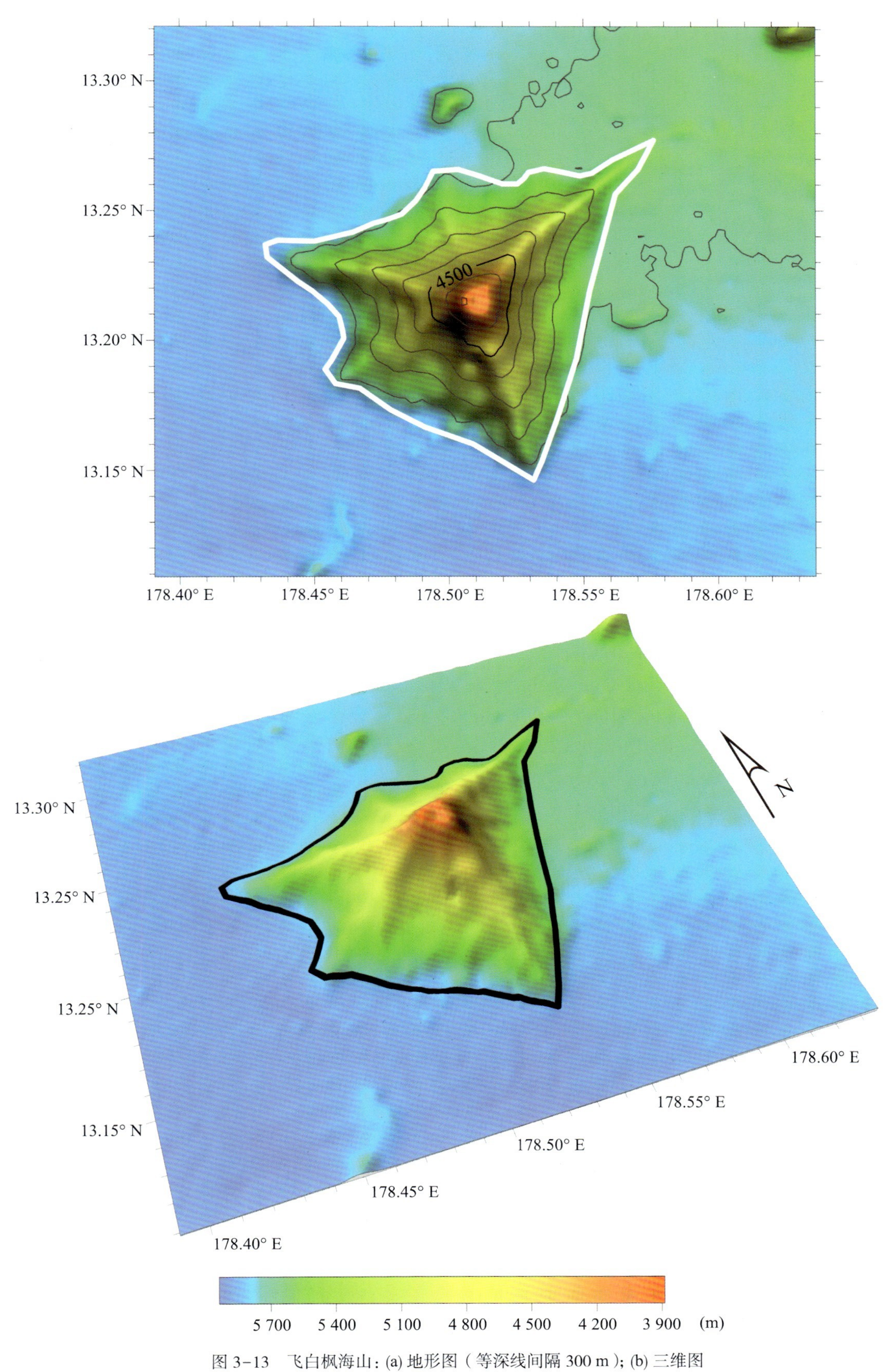

图 3-13 飞白枫海山：(a) 地形图（等深线间隔 300 m）；(b) 三维图

Fig.3-13 Feibaifeng Seamount: (a) Bathymetric map (Contours are in 300 m); (b) 3-D topographic map

3.2.13 鞘群平顶山群 Qiaoqun Guyots

中文名称 Chinese Name	Qiàoqún Píngdǐngshānqún 鞘群平顶山群		
英文名称 English Name	Qiaoqun Guyots		
地理区域 / Location	中北太平洋 The Central North Pacific Ocean		
特征点坐标 Coordinate	13°21.4′N 179°03.7′E	长度 / Length	74 km
		宽度 / Width	71 km
水深 / Depth	1 731~5 670 m	高差 / Total Relief	3 939 m
发现情况 Discovery Facts	此平顶山群于 2019 年“向阳红 03”船在执行航次调查时发现。 The guyots were discovered in 2019 during the survey cruise carried out onboard the Chinese R/V *Xiang Yang Hong 03*.		
地形特征 Feature Description	鞘群平顶山群位于中北太平洋，勒内礁正南 389 km。从平顶至坡脚均呈辐射状发育多条山脊。最高处位于北部规模较大的山体，水深 1 731 m，南部两个小型平顶山自东向西分别排列。该平顶山群孤立于平坦的深海平原中，周边水深超过 5 500 m。 Located in the Central North Pacific Ocean, the guyots are 389 km south of the Rene Reef. Many ridges are distributed radially from the top of the guyots to the foot of the slope. The highest point is located in a relatively large mountain body in the north, with a depth of 1,731 m. In the south, two small guyots extend from east to west respectively. The guyots are isolated on the flat abyssal plain, with a depth of more than 5,500 m in the surroundings.		
专名释义 Reason for Choice of Name	该地区新发现的海底地理实体均以太平洋生物群组化方法命名。鞘群，“*Epizoanthus*”种属的中文名称。“蛟龙”号和“发现”号潜水器在西太平洋海域发现鞘群海葵，俯视该平顶山群，形似鞘群海葵，故以鞘群命名。 The newly discovered undersea features in this sea area are all named after groups of the Pacific Ocean species. Qiaoqun is the Chinese name for the species taxa “*Epizoanthus*” family. The submersibles “Jiaolong” and “Faxian” have discovered Qiaoqun sea anemones in the Pacific Ocean. The undersea features look like Qiaoqun sea anemones from the top view. So the features are named Qiaoqun Guyots.		

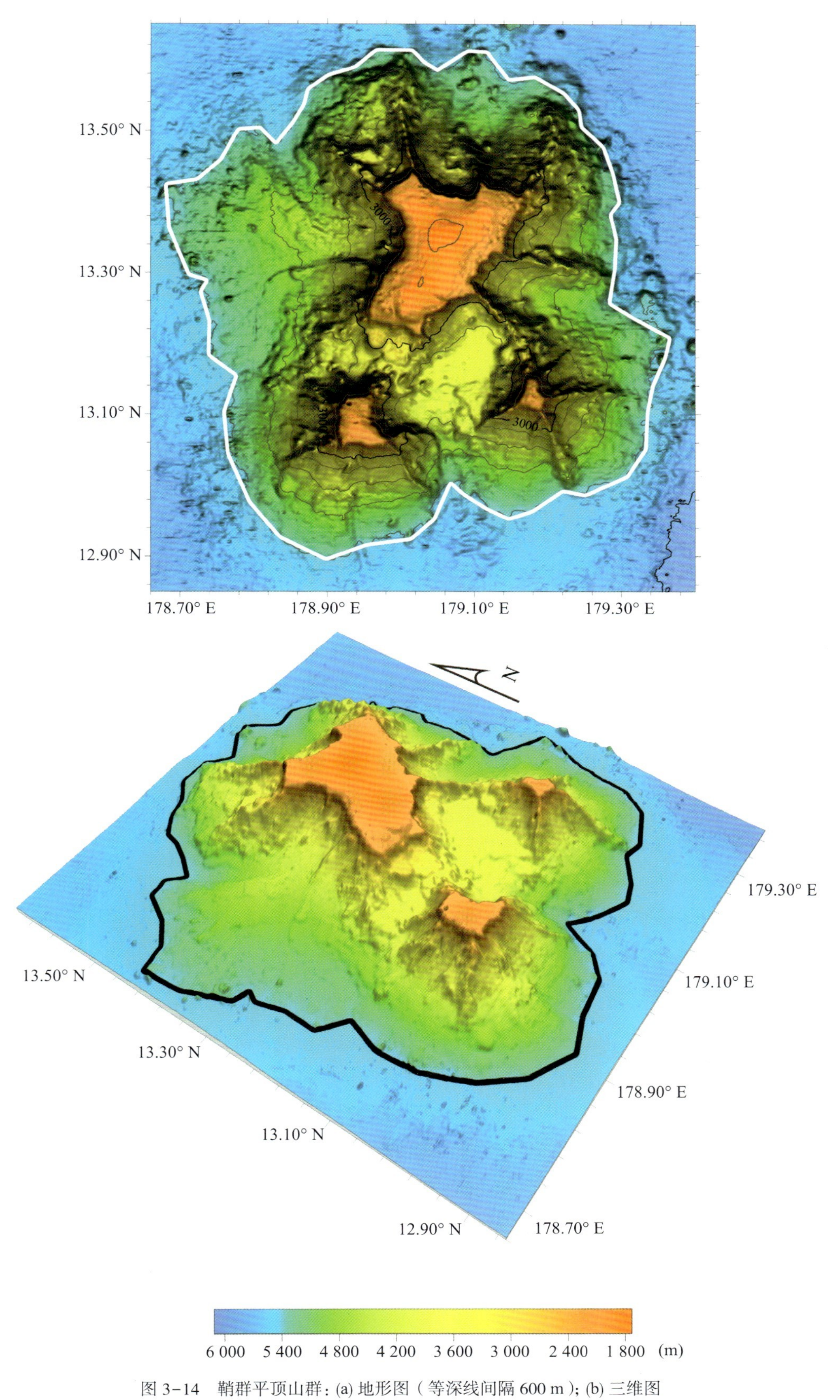

图 3-14 鞘群平顶山群：(a) 地形图（等深线间隔 600 m）；(b) 三维图
Fig.3-14 Qiaoqun Guyots: (a) Bathymetric map (Contours are in 600 m); (b) 3-D topographic map

3.2.14 灯笼鱼海山 Denglongyu Seamount

中文名称 Chinese Name	Dēnglongyú Hǎishān 灯笼鱼海山		
英文名称 English Name	Denglongyu Seamount		
地理区域 / Location	中北太平洋 The Central North Pacific Ocean		
特征点坐标 Coordinate	14°31.7′N 179°43.1′E	长度 / Length	81 km
		宽度 / Width	31 km
水深 / Depth	1 675~5 417 m	高差 / Total Relief	3 742 m
发现情况 Discovery Facts	此海山于 2019 年"向阳红 14"船在执行航次调查时发现。 The seamount was discovered in 2019 during the survey cruise carried out onboard the Chinese R/V *Xiang Yang Hong 14*.		
地形特征 Feature Description	灯笼鱼海山位于中北太平洋，勒内礁东南 318 km，整体呈 NE-SW 走向，长 81 km，宽 31 km。该海山由两个相连的山峰组成，最高处位于东北部规模较大的山峰，水深约 1 675 m。该海山周围为地形相对平坦的深海平原，平均水深超过 5 300 m。 Located in the Central North Pacific Ocean, the seamount is 318 km southeast of the Rene Reef. Extending in the NE-SW direction as a whole, it is 81 km long and 31 km wide. The seamount is composed of two connected peaks. The highest point is located in a relatively large peak in the northeast, with a depth of about 1,675 m. Around the seamount is a relatively flat abyssal plain with an average depth of over 5,300 m.		
专名释义 Reason for Choice of Name	灯笼鱼，英文名为"Anglerfish"，拉丁文名为"*Lophiiformes*"，一般生活在 500~5 000 m 的深海，因头顶突出诱饵吸引猎物而得名。俯视该海山形似一只游向东北方向的灯笼鱼，故以此命名。 Denglongyu is the Chinese name for the anglerfish, or *Lophiiformes* in Latin. The anglerfish generally live in the deep sea of 500~5,000 m with lure projecting from head to attract prey. The feature looks like an anglerfish swimming to the northeast from the top view. So the feature is named Denglongyu Seamount.		

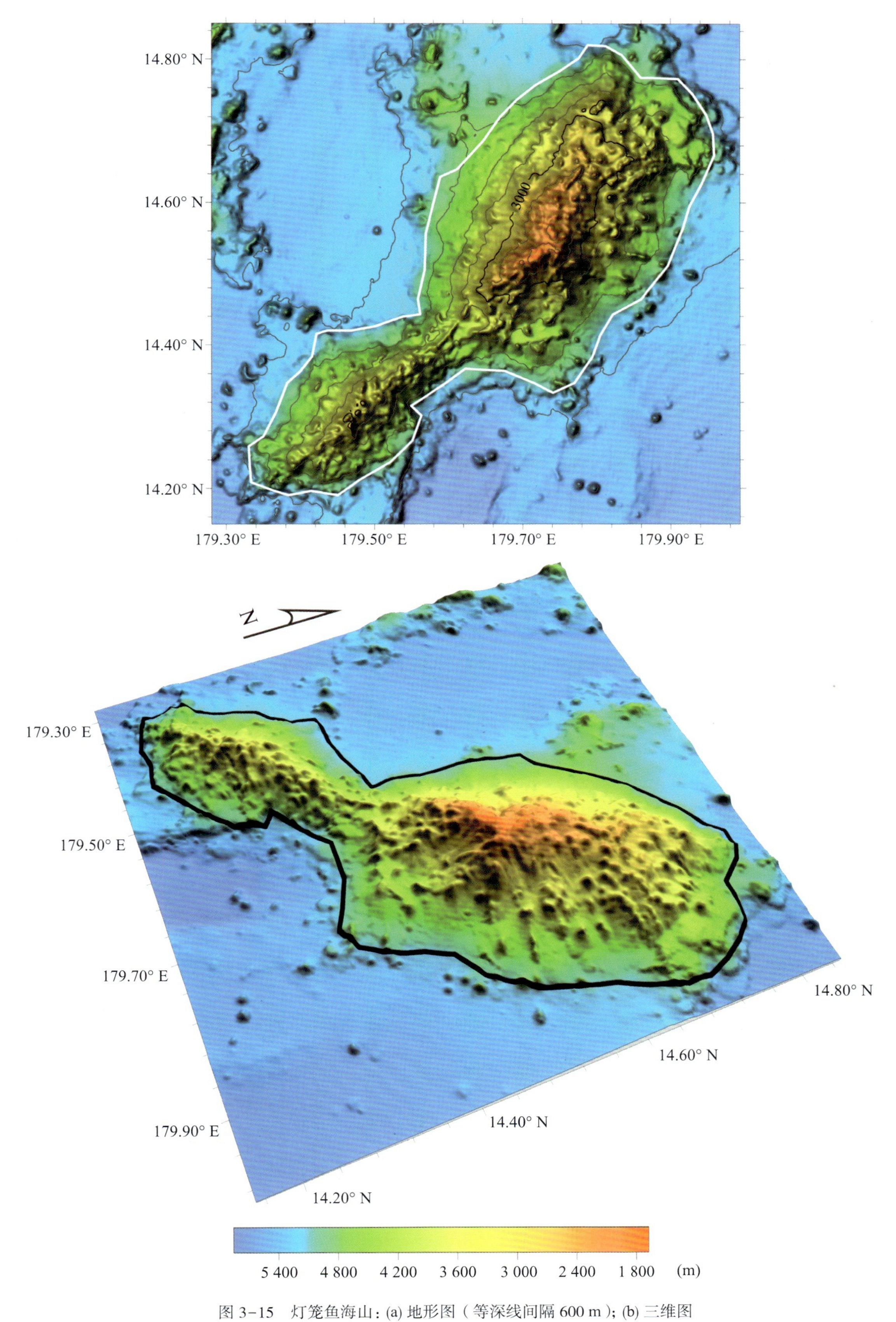

图 3-15 灯笼鱼海山：(a) 地形图（等深线间隔 600 m）；(b) 三维图

Fig.3-15 Denglongyu Seamount: (a) Bathymetric map (Contours are in 600 m); (b) 3-D topographic map

3.2.15 梅花参海山 Meihuashen Seamount

中文名称 Chinese Name	Méihuāshēn Hǎishān 梅花参海山		
英文名称 English Name	Meihuashen Seamount		
地理区域 / Location	中北太平洋 The Central North Pacific Ocean		
特征点坐标 Coordinate	13°23.4′N 179°37.3′E	长度 / Length	18 km
		宽度 / Width	16 km
水深 / Depth	3 547~5 846 m	高差 / Total Relief	2 299 m
发现情况 Discovery Facts	此海山于 2019 年“向阳红 03”船在执行航次调查时发现。 The seamount was discovered in 2019 during the survey cruise carried out onboard the Chinese R/V *Xiang Yang Hong 03*.		
地形特征 Feature Description	梅花参海山位于中北太平洋，勒内礁东南 391 km。该海山大致呈等维分布，长 18 km，宽 16 km，最高处位于海山中心，水深约 3 547 m。该海山东部紧邻一座大型海山，其余三个方向为地形平坦的深海平原，平均水深超过 5 700 m。 Located in the Central North Pacific Ocean, the seamount is 391 km southeast of the Rene Reef. Being generally equidimensional in plan, it is 18 km long and 16 km wide. The highest point is located in the center of the seamount, with a depth of about 3,547 m. It is adjacent to a large seamount in the east, and there are flat abyssal plains with an average depth of more than 5,700 m in the other three directions (south, north and west) of the seamount.		
专名释义 Reason for Choice of Name	该地区新发现的海底地理实体均以太平洋生物群组化方法命名。梅花参，“*Thelenota*”种属的中文名称。“蛟龙”号和“发现”号潜水器在太平洋海域发现了梅花参。故以梅花参命名。 The newly discovered undersea features in this area are all named after groups of the Pacific Ocean species. Meihuashen is the Chinese name for the species taxa “*Thelenota*” genus. The submersibles “Jiaolong” and “Faxian” have discovered Meihuashen in the Pacific Ocean. So the undersea feature is named Meihuashen Seamount.		

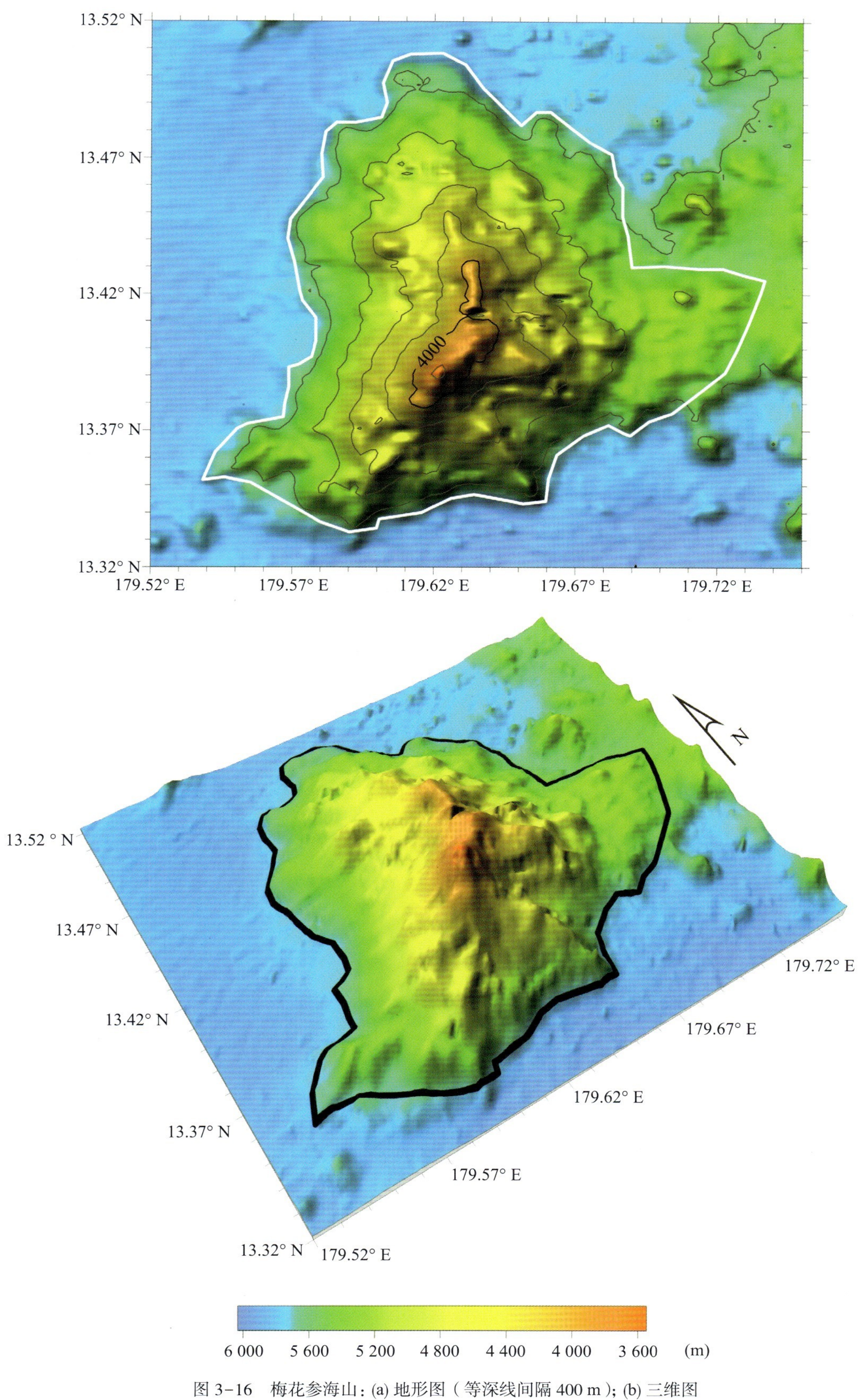

图 3-16 梅花参海山：(a) 地形图（等深线间隔 400 m）；(b) 三维图

Fig.3-16 Meihuashen Seamount: (a) Bathymetric map (Contours are in 400 m); (b) 3-D topographic map

东北太平洋海底地理实体

Chapter IV

The Undersea Features in the Northeast Pacific Ocean

4.1 地形地貌概况

东北太平洋东岸大陆架狭窄，最窄的大陆架宽 18~20 km，东岸海岸线平直，边缘部分发育多种地貌形态，包括阿拉斯加湾海山区和中美洲海槽等。北美山系直逼东岸海岸线，其中科迪勒拉山系是世界上火山活动最剧烈的地带之一，地质活动多发。帝王海山链、夏威夷海脊和中太平洋海山群将深水区分为西北太平洋海盆、东北太平洋海盆和中太平洋海盆。东北太平洋海盆作为世界上最大的海盆，面积约 4 874 km^2，相当于整个太平洋面积的 26.9%，其北至阿留申海槽，南至南太平洋海丘，南北长 8 900 km；西至帝王海山链和夏威夷海脊，东侧是北美大陆架，东西宽约 4 200 km。东北太平洋海盆深邃而广阔，水深多为 5 000~5 500 m，最大水深 7 168 m，盆底地形平缓，底部发育了一些小海山，其直径多为 12~15 km；自西向东发育多条数千千米的纬向断裂带，宽为 100~200 km，如门多西诺断裂带、穆雷断裂带和莫罗凯断裂带等，共计 14 条。

Section 4.1 Overview of the topography

The continental shelf on the east coast of the Northeast Pacific Ocean is relatively narrow, with the narrowest shelf being about18~20 km wide. The coastline on the east coast is straight, with the development of a variety of geographic forms at the edges, including the Alaska Bay Seamount area and the Central American Trough. The North American mountain range extends directly to the east coastline, and its Cordillera mountain range is one of the most volcanically active areas in the world, with frequent geological activities. The Emperor Seamount chain, Hawaiian Ridge and the Central Pacific Seamounts divide the deep sea area into the Northwest Pacific Basin, the Northeast Pacific Basin and the Central Pacific Basin. As the largest basin in the world, the Northeast Pacific Basin covers an area of 4,874 km^2, equivalent to 26.9% of the Pacific Ocean. It stretches from the Aleutian Trough in the north to the South Pacific Sea Hill in the south, with a length of 8,900 km from north to south. To the west is the Emperor Seamount Chain and the Hawaiian Ridge, and to the east is the North American continental shelf, which is about 4,200 km wide. The Northeast Pacific Basin is deep and vast, with depths of 5,000~5,500 m and a maximum depth of 7,168 m. The basin floor topography is smooth, and some small seamounts have developed at the bottom with diameters mostly of 12~15 km. From west to east, there are more than 14 latitudinal fracture zones with the lengths being about several thousand km and the widths ranging from 100 km to 200 km, such as Mendocino Fracture Zone, Murray Fracture Zone and Molokai Fracture Zone.

4.2 东北太平洋地理实体命名

考虑到命名的海底地理实体与SCUFN通过的音乐家海山群距离较近，我国对东北太平洋的海底地理实体进行命名时，主要沿用音乐家相关命名体系及中国乐器相关命名体系群组化命名方法。这些海底地理实体主要分布在门多西诺断裂带以北，东北太平洋海盆东北部。

新命名的海底地理实体共28个，其中海山及海山群17个，为钟仪海山、师涓海山、师旷海山、师襄海山、钟子期海山、伯牙海山、桓谭海山、蔡邕海山、万宝常海山群、苏祇婆海山、李龟年海山、段善本海山、姜夔海山群、雷海青海山群、朱载堉海山、魏良辅海山、船锚海山；海丘及海丘群3个，为葫芦埙海丘群、编钟海丘群、瑶琴海丘；圆丘3个，为阮咸圆丘、铜钹圆丘、铜锣圆丘；海脊及海脊群4个，为箜篌海脊群、鸿鹄海脊、梆笛海脊、曲笛海脊；海渊1个，为芦笙海渊。

Section 4.2 Undersea features in the Northeast Pacific Ocean

The undersea features in this sea area in the Northeast Pacific Ocean are all named following grouping naming system related to Chinese musicians and musical instruments, considering that the undersea features are adjacent to the Musicians Seamounts selected by SCUFN. These undersea features are mainly distributed in the north of the Mendocino Fracture Zone and northeast of the Northeast Pacific Basin.

We have named a total of 28 undersea features newly discovered in the Northeast Pacific Ocean, including 17 seamounts with 3 relatively gathering seamounts (groups) : Zhongyi Seamount, Shijuan Seamount, Shikuang Seamount, Shixiang Seamount, Zhongziqi Seamount, Boya Seamount, Huantan Seamount, Caiyong Seamount, Wanbaochang Seamounts (Group), Suzhipo

Seamount, Liguinian Seamount, Duanshanben Seamount, Jiangkui Seamounts (Group), Leihaiqing Seamounts (Group), Zhuzaiyu Seamount, Weiliangfu Seamount and Chuanmao Seamount; 1 hill and 2 relatively gathering hills (groups): Bianzhong Hills (Group), Huluxun Hills (Group), Yaoqin Hill; 3 knolls : Ruanxian Knoll, Tongbo Knoll, Tongluo Knoll; 3 ridges and 1 relatively gathering ridges (group) : Konghou Ridges (Group), Bangdi Ridge, Honghu Ridge and Qudi Ridge; 1 deep : Lusheng Deep.

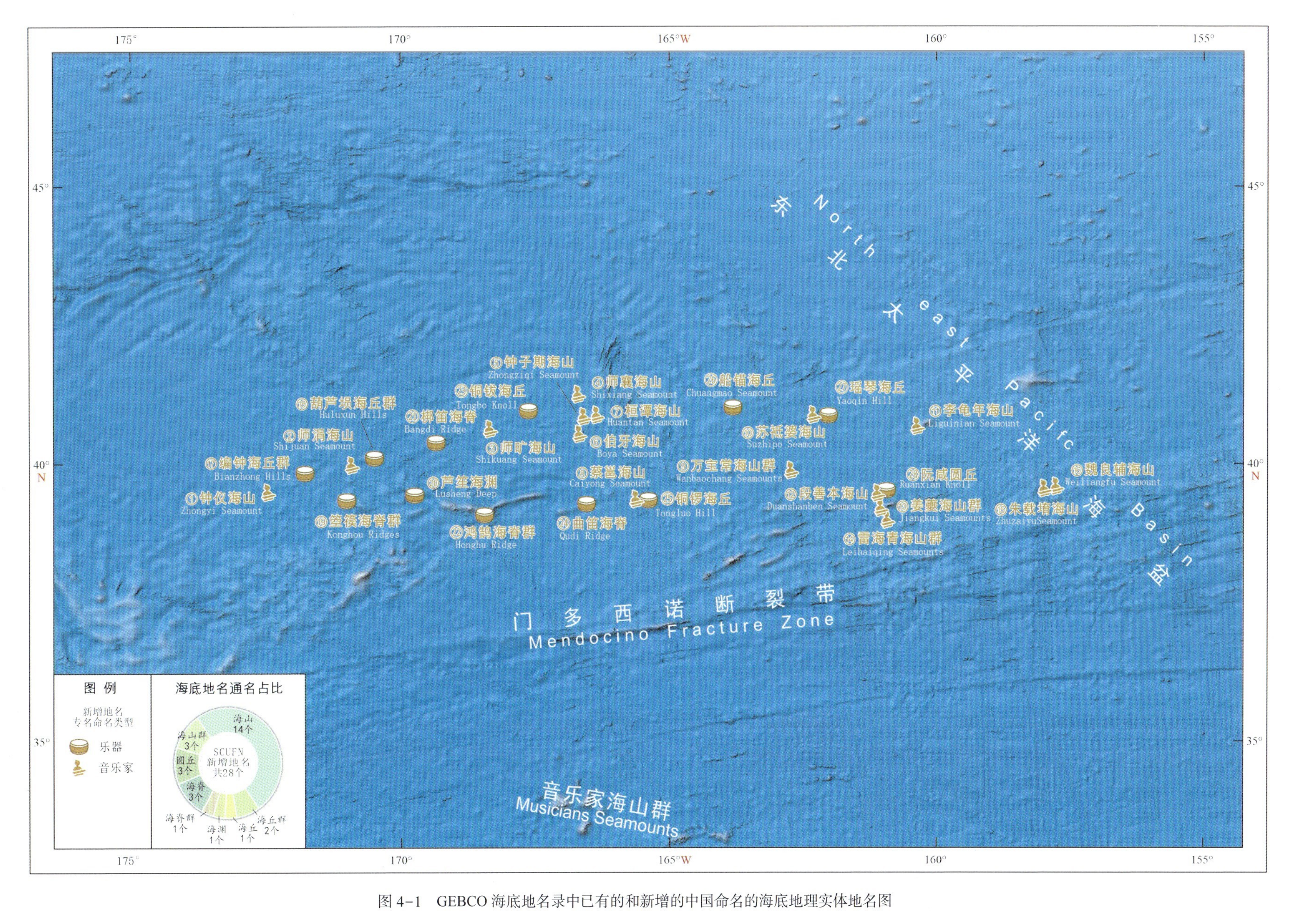

图 4-1 GEBCO 海底地名录中已有的和新增的中国命名的海底地理实体地名图

Fig.4-1 Map of the undersea feature names proposed by China and already included in GEBCO Gazetteer and these newly proposed by China

1	钟仪海山	Zhongyi Seamount	39°29.7′N 172°26.4′W
2	师涓海山	Shijuan Seamount	39°59.1′N 170°53.2′W
3	师旷海山	Shikuang Seamount	40°38.3′N 168°20.3′W
4	师襄海山	Shixiang Seamount	41°15.6′N 166°41.4′W
5	钟子期海山	Zhongziqi Seamount	40°51.4′N 166°35.3′W
6	伯牙海山	Boya Seamount	40°32.7′N 166°40.3′W
7	桓谭海山	Huantan Seamount	40°53.4′N 166°20.5′W
8	蔡邕海山	Caiyong Seamount	39°22.0′N 165°36.5′W
9	万宝常海山群	Wanbaochang Seamounts	39°56.8′N 162°40.6′W 39°53.2′N 162°42.4′W 39°48.9′N 162°48.7′W
10	苏祇婆海山	Suzhipo Seamount	40°53.6′N 162°17.2′W
11	李龟年海山	Liguinian Seamount	40°40.9′N 160°20.0′W
12	段善本海山	Duanshanben Seamount	39°28.4′N 161°05.1′W
13	姜夔海山群	Jiangkui Seamounts	39°14.9′N 160°50.4′W 39°12.8′N 161°01.2′W 39°12.0′N 161°07.3′W
14	雷海青海山群	Leihaiqing Seamounts	39°00.0′N 160°53.9′W 39°01.3′N 160°46.7′W

15	朱载堉海山	Zhuzaiyu Seamount	39°33.5′N 157°58.2′W
16	魏良辅海山	Weiliangfu Seamount	39°36.0′N 157°45.2′W
17	编钟海丘群	Bianzhong Hills	39°54.0′N 171°36.7′W 39°49.9′N 171°45.8′W
18	葫芦埙海丘群	Huluxun Hills	40°05.2′N 170°38.9′W 40°03.8′N 170°36.9′W 40°06.6′N 170°29.0′W
19	箜篌海脊群	Konghou Ridges	39°12.2′N 170°41.3′W 39°20.5′N 170°59.6′W 39°18.7′N 171°01.3′W
20	芦笙海渊	Lusheng Deep	39°26.2′N 169°45.0′W
21	梆笛海脊	Bangdi Ridge	40°23.3′N 169°21.3′W
22	鸿鹄海脊	Honghu Ridge	39°04.9′N 168°27.1′W
23	铜钹海丘	Tongbo Hill	40°56.9′N 167°38.2′W
24	曲笛海脊	Qudi Ridge	39°16.9′N 166°32.7′W
25	铜锣圆丘	Tongluo Knoll	39°21.0′N 165°23.6′W
26	船锚海山	Chuanmao Seamount	41°00.9′N 163°48.8′W
27	瑶琴海丘	Yaoqin Hill	40°52.6′N 162°00.9′W
28	阮咸圆丘	Ruanxian Knoll	39°31.4′N 160°55.7′W

4.2.1 钟仪海山 Zhongyi Seamount

<table>
<tr><td>中文名称
Chinese Name</td><td colspan="3">Zhōngyí Hǎishān
钟仪海山</td></tr>
<tr><td>英文名称
English Name</td><td colspan="3">Zhongyi Seamount</td></tr>
<tr><td>地理区域 / Location</td><td colspan="3">东北太平洋 The Northeast Pacific Ocean</td></tr>
<tr><td rowspan="2">特征点坐标
Coordinate</td><td rowspan="2">39°29.7′N 172°26.4′W</td><td>长度 / Length</td><td>15.5 km</td></tr>
<tr><td>宽度 / Width</td><td>14.5 km</td></tr>
<tr><td>水深 / Depth</td><td>4 376~5 874 m</td><td>高差 / Total Relief</td><td>1 498 m</td></tr>
<tr><td>发现情况
Discovery Facts</td><td colspan="3">此海山于 2019 年“向阳红 19”船在执行航次调查时发现。
The seamount was discovered in 2019 during the survey cruise carried out onboard the Chinese R/V Xiang Yang Hong 19.</td></tr>
<tr><td>地形特征
Feature Description</td><td colspan="3">钟仪海山位于东北太平洋，音乐家海山群西北方向 1 100 km。整体呈 E-W 走向，长 15.5 km，宽 14.5 km；顶部平坦，最高处位于海山东南部，水深约 4 376 m；峰顶东侧存在一个小型坳陷，高差超过 300 m，北侧山麓发育规模不等的小型海丘。该海山孤立于深海平原中，周围地形较为平坦，但密布小型沟槽，发育规模较小的深海海丘，周边平均水深超过 5 800 m。
Located in the Northeast Pacific Ocean, the seamount is 1,100 km northwest of the Musicians Seamounts. It extends in the E-W direction as a whole, and is 15.5 km long and 14.5 km wide. Being flat at the top, it is the highest in the southeast of the flat top, with a depth of about 4,376 m. In the east of the peak, there is a small depression, which has a total relief of more than 300 m. At the northern foot, there are many small hills of different sizes. With relatively flat surroundings, the seamount stands isolated on the abyssal plain, but is densely covered with small trenches extending in the W-S direction. There are also small deep-sea hills, with an average depth of more than 5,800 m in the surroundings.</td></tr>
<tr><td>专名释义
Reason for Choice of Name</td><td colspan="3">该海山邻近 SCUFN 已收录的音乐家海山群，沿用音乐家相关命名体系群组化方法命名。钟仪，春秋时期楚国人，楚国公族，是有史书记载最早的古琴演奏家，世代为宫廷琴师。故以钟仪命名。
Being adjacent to the Musicians Seamounts that have been selected by SCUFN, the undersea features in this sea area are all named following grouping naming system related to musicians. There was a famous musician named Zhongyi, a native and important musician of the State of Chu during the Spring and Autumn Period of China. He was the earliest Guqin (a seven-stringed plucked instrument in some ways similar to the zither) player according to the historical records, and his ancestors were court Guqin players for many generations. Hence, the feature is named Zhongyi Seamount after him.</td></tr>
</table>

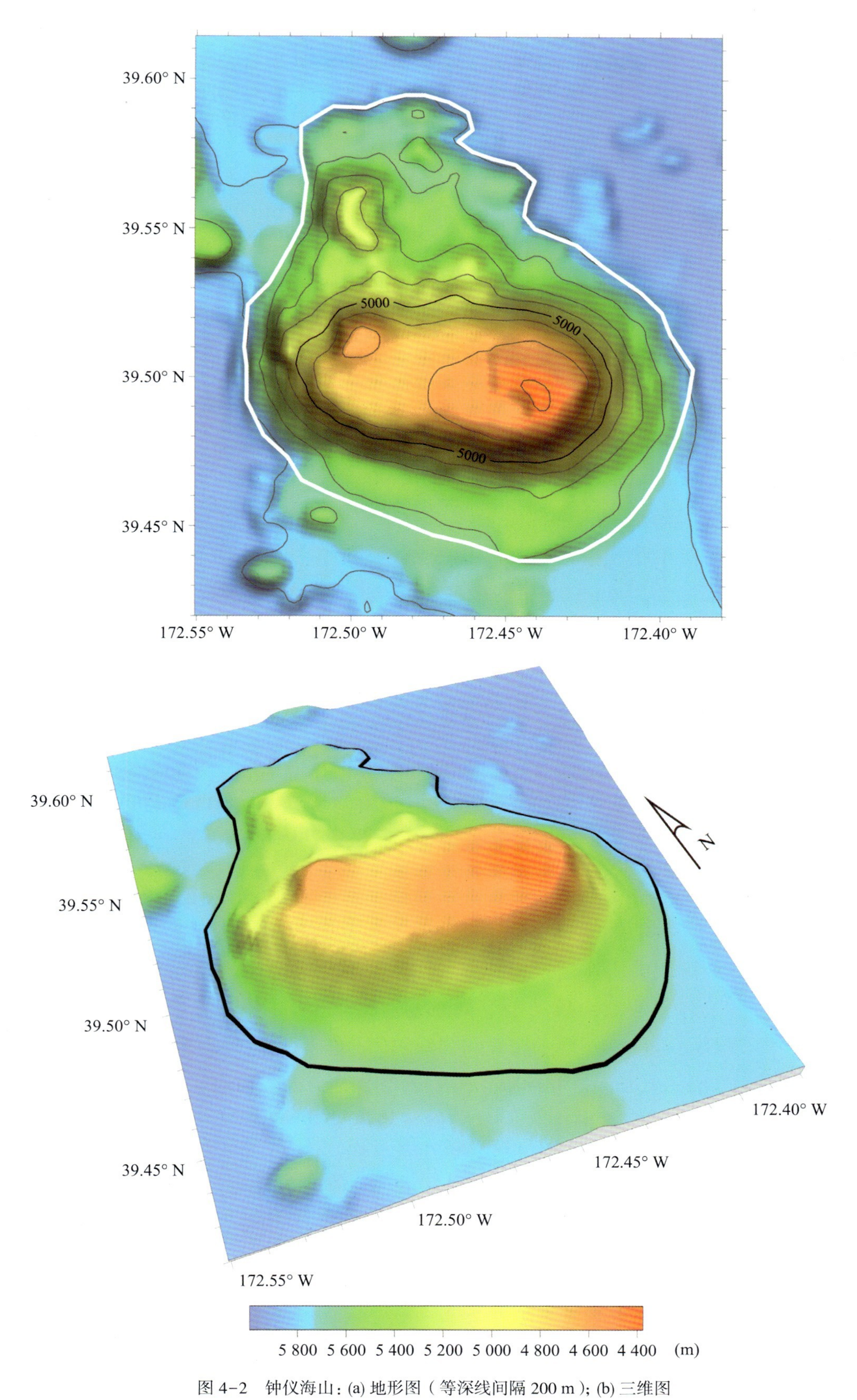

图 4-2 钟仪海山：(a) 地形图（等深线间隔 200 m）；(b) 三维图

Fig.4-2 Zhongyi Seamount: (a) Bathymetric map (Contours are in 200 m); (b) 3-D topographic map

4.2.2 师涓海山 Shijuan Seamount

<table>
<tr><td>中文名称
Chinese Name</td><td colspan="3">Shījuān Hǎishān
师涓海山</td></tr>
<tr><td>英文名称
English Name</td><td colspan="3">Shijuan Seamount</td></tr>
<tr><td>地理区域 / Location</td><td colspan="3">东北太平洋 The Northeast Pacific Ocean</td></tr>
<tr><td rowspan="2">特征点坐标
Coordinate</td><td rowspan="2">39°59.1′N 170°53.2′W</td><td>长度 / Length</td><td>14.5 km</td></tr>
<tr><td>宽度 / Width</td><td>12 km</td></tr>
<tr><td>水深 / Depth</td><td>4 657~5 748 m</td><td>高差 / Total Relief</td><td>1 091 m</td></tr>
<tr><td>发现情况
Discovery Facts</td><td colspan="3">此海山于 2019 年“向阳红 19”船在执行航次调查时发现。
The seamount was discovered in 2019 during the survey cruise carried out onboard the Chinese R/V Xiang Yang Hong 19.</td></tr>
<tr><td>地形特征
Feature Description</td><td colspan="3">师涓海山位于东北太平洋，音乐家海山群西北方向 1 127 km。该海山大致呈等维状展布，东西长 14.5 km，南北宽 12 km，最高处位于海山中部峰顶，水深约为 4 657 m；海山西、南两侧坡度较缓，东北侧坡度较大。该海山孤立于深海平原中，周围地形较为平坦，但密布小型沟槽，周边平均水深超过 5 800 m。
Located in the Northeast Pacific Ocean, the seamount is 1,127 km northwest of the Musicians Seamounts. Being generally equidimensional in plan, it is 14.5 km long from west to east and 12 km wide from south to north. The highest point is located at the peak in the middle of the seamount, with a depth of about 4,657 km. The slopes of the seamount in the west and south are relatively gentle whereas the slope in the northeast is relatively steep. With relatively flat surroundings, the seamount stands isolated on the abyssal plain, but is densely covered with small trenches, with an average depth exceeding 5,800 m in the surroundings.</td></tr>
<tr><td>专名释义
Reason for Choice of Name</td><td colspan="3">该区域沿用音乐家相关命名体系群组化方法命名。师涓，我国春秋时期卫国著名音乐家，活动于卫灵公（公元前 534—公元前 492）在位期间，以善弹琴而著称，并善于搜集和弹奏民间乐曲。故以师涓命名。
The features in this sea area are all named following grouping naming system related to musicians. There was a famous musician named Shi Juan, a well-known musician in the State of Wei during the Spring and Autumn Period of China. He served the Reign of Duke Ling of Wey (534BC−492 BC). He was famous for playing the stringed instruments and was good at collecting and playing folk music. So the feature is named Shijuan Seamount after him.</td></tr>
</table>

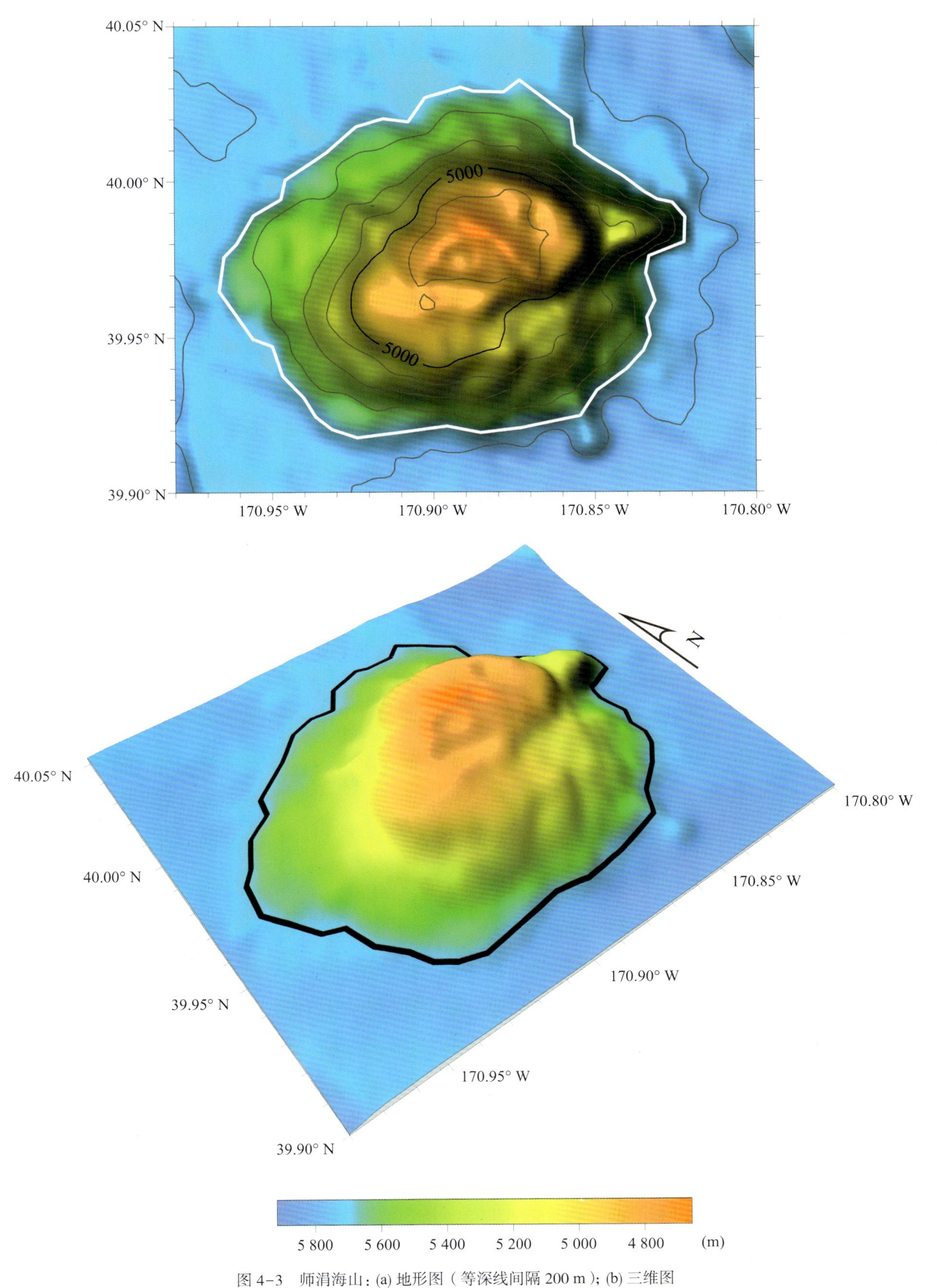

图 4-3 师涓海山：(a) 地形图（等深线间隔 200 m）；(b) 三维图

Fig.4-3 Shijuan Seamount: (a) Bathymetric map (Contours are in 200 m); (b) 3-D topographic map

4.2.3 师旷海山 Shikuang Seamount

<table>
<tr><td>中文名称
Chinese Name</td><td colspan="3">Shīkuàng Hǎishān
师旷海山</td></tr>
<tr><td>英文名称
English Name</td><td colspan="3">Shikuang Seamount</td></tr>
<tr><td>地理区域 / Location</td><td colspan="3">东北太平洋 The Northeast Pacific Ocean</td></tr>
<tr><td rowspan="2">特征点坐标
Coordinate</td><td rowspan="2">40°38.3′N 168°20.3′W</td><td>长度 / Length</td><td>17 km</td></tr>
<tr><td>宽度 / Width</td><td>11 km</td></tr>
<tr><td>水深 / Depth</td><td>4 501~5 922 m</td><td>高差 / Total Relief</td><td>1 421 m</td></tr>
<tr><td>发现情况
Discovery Facts</td><td colspan="3">此海山于 2019 年“向阳红 06”船在执行航次调查时发现。
The seamount was discovered in 2019 during the survey cruise carried out onboard the Chinese R/V Xiang Yang Hong 06.</td></tr>
<tr><td>地形特征
Feature Description</td><td colspan="3">师旷海山位于东北太平洋，音乐家海山群西北方向 1 113 km。该海山呈 E-W 走向，主体呈圆形，东西长约 17 km，南北宽约 11 km；峰顶存在一个圆形凹陷，与周边高差超过 200 m。该海山东、西两侧广泛分布小型凸起，与海山共同呈 E-W 走向分布；南北两侧为地形平坦的深海平原，平均水深约为 5 800 m。
Located in the Northeast Pacific Ocean, the seamount is 1,113 km northwest of the Musicians Seamounts. Being rounded in the main body, it extends in the E-W direction, and is about 17 km long from west to east, and about 11 km wide from south to north. There is a rounded depression at the peak, with a total relief of more than 200 m over the surrounding areas. Small protrusions are widely distributed in the east and west, which extends in the same E-W direction as the seamount. There are flat abyssal plains in the north and south of the seamount, with an average depth of about 5,800 m.</td></tr>
<tr><td>专名释义
Reason for Choice of Name</td><td colspan="3">该区域沿用音乐家相关命名体系群组化方法命名。师旷，字子野，平阳（今山东省新泰市南师店）人，先秦著名音乐大师，春秋时期晋国晋悼公、晋平公时大臣、太宰、宫廷掌乐大师，博学多才，尤精音律，善弹琴，辨音力极强，古人称乐圣。故以师旷命名。
The features in this sea areas are all named following grouping naming system related to musicians. There was a famous musician named Shi Kuang, with the courtesy name Ziye, born in Pingyang (now Nanshidian, Xintai City, Shandong Province). He was a famous musician during the pre-Qin Period of China, and assumed important positions including prime minister, and person in charge of the court music during the Reigns of Duke Dao and Duke Ping in the State of Jin during the Spring and Autumn Period of China. Being versatile, he was a master of music, and had a strong ability to play stringed instruments and distinguish sounds. Therefore, he was called Sage of Music by the ancients. So the feature is named Shikuang Seamount after him.</td></tr>
</table>

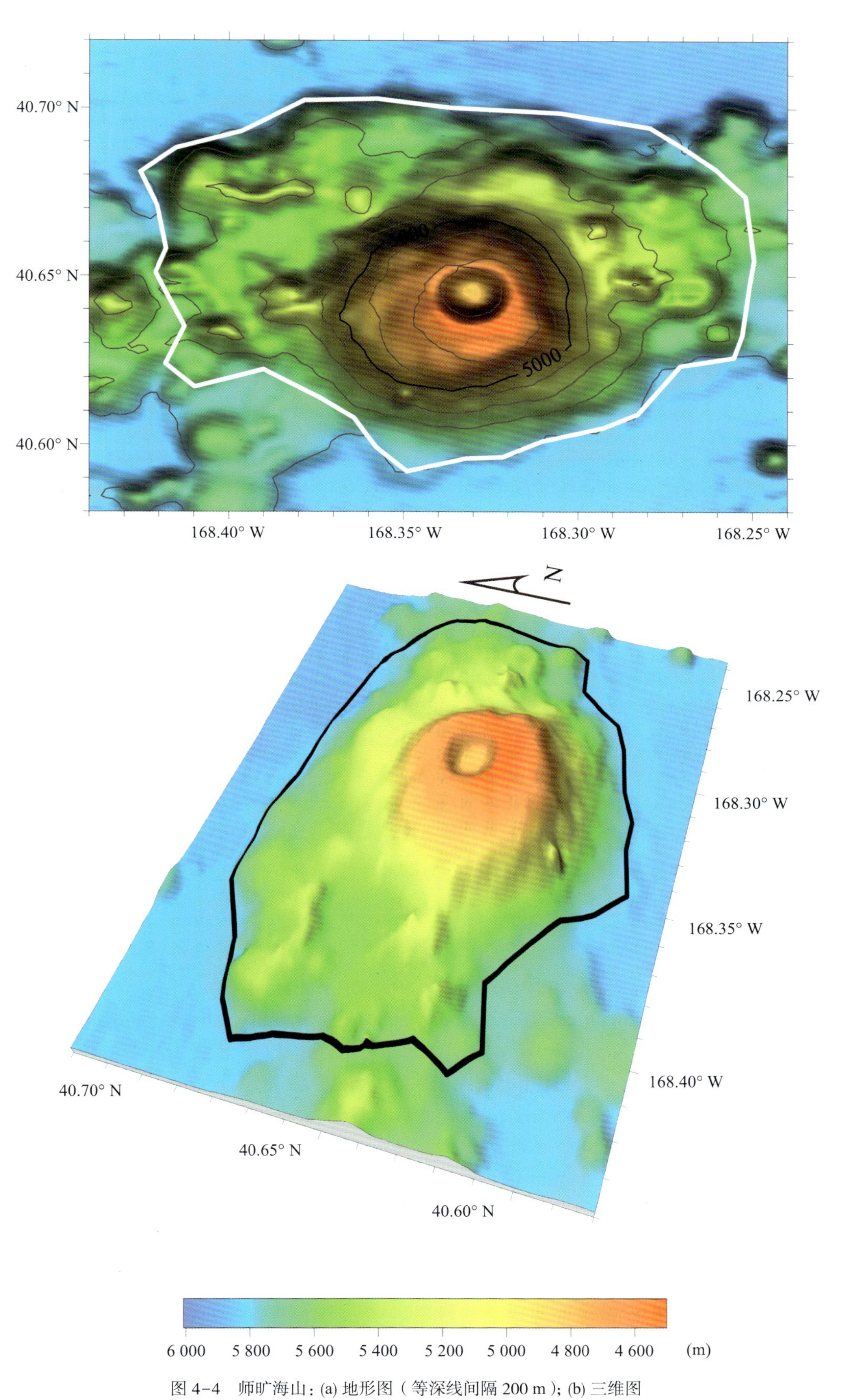

图 4-4 师旷海山：(a) 地形图（等深线间隔 200 m）；(b) 三维图

Fig.4-4 Shikuang Seamount: (a) Bathymetric map (Contours are in 200 m); (b) 3-D topographic map

4.2.4 师襄海山 Shixiang Seamount

<table>
<tr><td>中文名称
Chinese Name</td><td colspan="3">Shīxiāng Hǎishān
师襄海山</td></tr>
<tr><td>英文名称
English Name</td><td colspan="3">Shixiang Seamount</td></tr>
<tr><td>地理区域 / Location</td><td colspan="3">东北太平洋 The Northeast Pacific Ocean</td></tr>
<tr><td rowspan="2">特征点坐标
Coordinate</td><td rowspan="2">41°15.6′N 166°41.4′W</td><td>长度 / Length</td><td>15 km</td></tr>
<tr><td>宽度 / Width</td><td>10 km</td></tr>
<tr><td>水深 / Depth</td><td>4 175~5 569 m</td><td>高差 / Total Relief</td><td>1 394 m</td></tr>
<tr><td>发现情况
Discovery Facts</td><td colspan="3">此海山于 2019 “向阳红 06”船在执行航次调查时发现。
The seamount was discovered in 2019 during the survey cruise carried out onboard the Chinese R/V Xiang Yang Hong 06.</td></tr>
<tr><td>地形特征
Feature Description</td><td colspan="3">师襄海山位于东北太平洋，音乐家海山群西北方向 1 141 km。该海山大致呈长方形，NE-SW 走向，东西长约 15 km，南北宽约 10 km，最高处位于海山西南部峰顶，水深约为 4 175 m。该海山孤立于地形较为平坦的深海平原，平均水深超过 5 500 m。
Located in the Northeast Pacific Ocean, the seamount is 1,141 km northwest of the Musicians Seamounts. Roughly in the shape of a rectangular, it extends in the NE-SW direction, and is about 15 km long from west to east, and about 10 km wide from south to north. The highest point is located at the peak of the seamount in the southwest, with a depth of about 4,175 m. The seamount is isolated on a relatively flat abyssal plain with an average depth of more than 5,500 m.</td></tr>
<tr><td>专名释义
Reason for Choice of Name</td><td colspan="3">该区域沿用音乐家相关命名体系群组化方法命名。师襄，亦称师襄子，我国春秋时期音乐大师，鲁国宫廷乐官，也有文献记载为卫国乐官。孔子曾向其学习弹琴，是圣人孔子的音乐老师。故以师襄命名。
The features in this sea area are all named following grouping naming system related to musicians. There was a famous musician named Shi Xiang, also known as Shi Xiangzi. He was a musician during the Spring and Autumn Period of China. He was a court music official of the State of Lu, or a music official of the State of Way according to some historical records. Confucius once learned to play instruments from him, and so he was the music teacher of the sage Confucius. So the feature is named Shixiang Seamount after him.</td></tr>
</table>

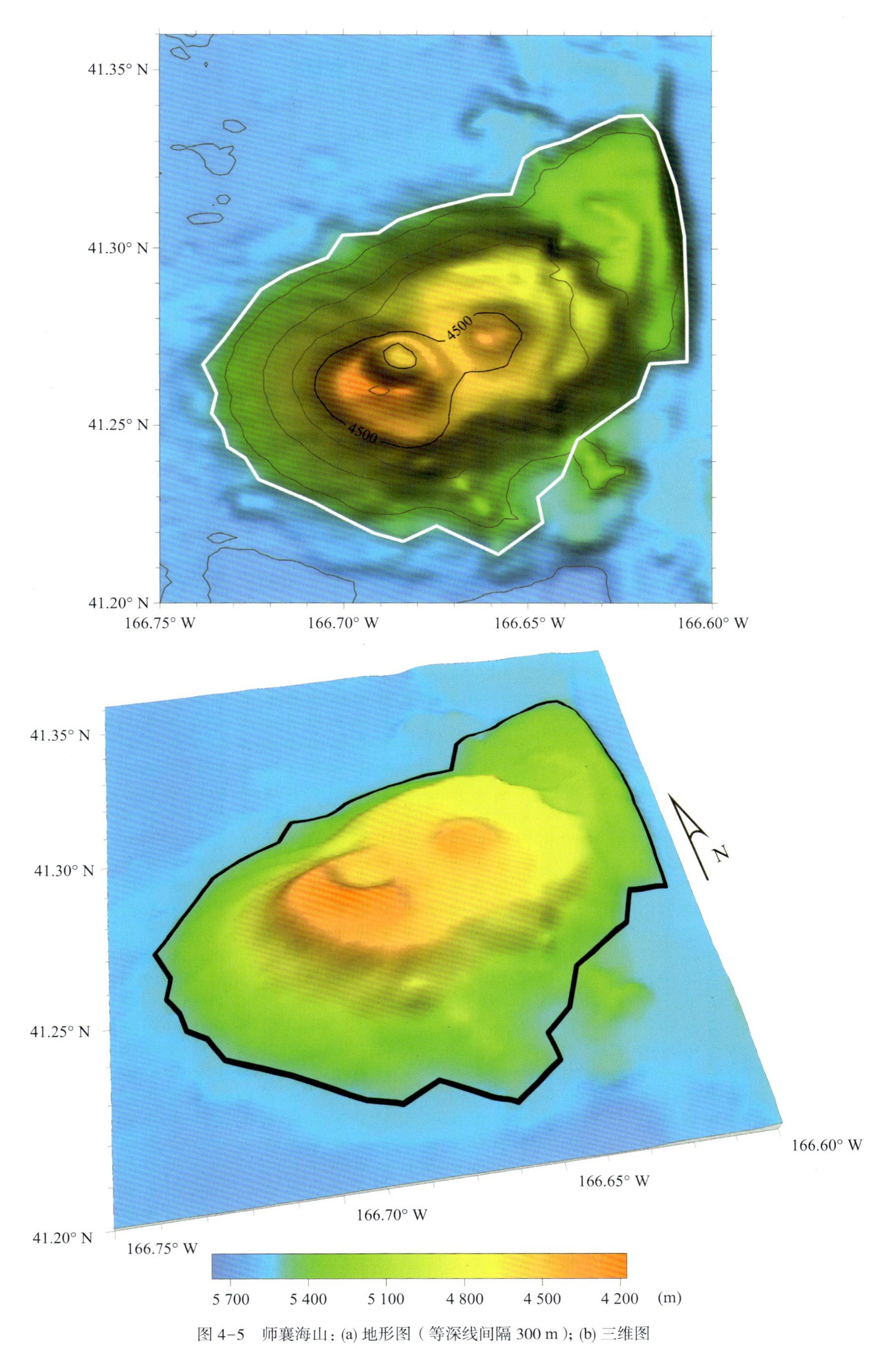

图 4-5 师襄海山：(a) 地形图（等深线间隔 300 m）；(b) 三维图

Fig.4-5 Shixiang Seamount: (a) Bathymetric map (Contours are in 300 m); (b) 3-D topographic map

4.2.5 钟子期海山 Zhongziqi Seamount

中文名称 **Chinese Name**	Zhōngzǐqī Hǎishān 钟子期海山		
英文名称 **English Name**	Zhongziqi Seamount		
地理区域 / Location	东北太平洋 The Northeast Pacific Ocean		
特征点坐标 **Coordinate**	40°51.4′N 166°35.3′W	**长度 / Length**	14.5 km
		宽度 / Width	12 km
水深 / Depth	4 156~5 652 m	**高差 / Total Relief**	1 496 m
发现情况 **Discovery Facts**	此海山于 2019 年“向阳红 06”船在执行航次调查时发现。 The seamount was discovered in 2019 during the survey cruise carried out onboard the Chinese R/V *Xiang Yang Hong 06.*		
地形特征 **Feature Description**	钟子期海山位于东北太平洋，音乐家海山群西北方向 1 081 km。该海山大致呈椭圆形，N–S 走向，南北长约 14.5 km，东西宽约 12 km，最高处位于海山中部，峰顶水深约为 4 156 m。该海山东侧地形较陡，西部相对平缓。 Located in the Northeast Pacific Ocean, the seamount is 1,081 km northwest of the Musicians Seamounts. Roughly in an oval shape, the seamount extends in the S-N direction, and is about 14.5 km long from south to north, and about 12 km wide from west to east. The highest point is located in the middle of the seamount, and the peak is about 4,156 m deep. The seamount is relatively steep in the east, and relatively flat in the west.		
专名释义 **Reason for Choice of Name**	该区域沿用音乐家相关命名体系群组化方法命名。钟子期，春秋战国时期楚国汉阳（今湖北省武汉市蔡甸区集贤村）人。《吕氏春秋》和《列子》中记载了伯牙与钟子期的故事，相传钟子期是一个戴斗笠、披蓑衣、背扁担、拿板斧的樵夫，他听到伯牙在汉江边鼓琴，感叹他高超的技艺，从此两人成了至交。该海山位于伯牙海山的北面，犹如钟子期席地而坐聆听伯牙“巍巍乎若高山，荡荡乎若流水”的琴声。故以钟子期命名。 The features in this sea area are all named following grouping naming system related to musicians. There was a famous musician named Zhong Ziqi who was a musician in the State of Chu during the Spring and Autumn and Warring States Periods of China. The story of Boya and Zhong Ziqi is recorded by *Mister Lv's Spring and Autumn Annals* and *Liezi*. According to the legend, Zhong Ziqi, a woodcutter who wore a bamboo hat and a straw rain cape, took a farming tool for carrying the load, and carried a broad axe, marveled at Boya's superior skills while the latter was playing the stringed instrument at the waterfront of Han River. Since then, they have been very intimate friends. To the north of Boya Seamount, this seamount looks like Zhong Ziqi sitting on the ground to watch Boya playing the music “as loft as the towering mountain, and as turbulent as the rushing water”. Therefore, the feature is named Zhongziqi Seamount after him.		

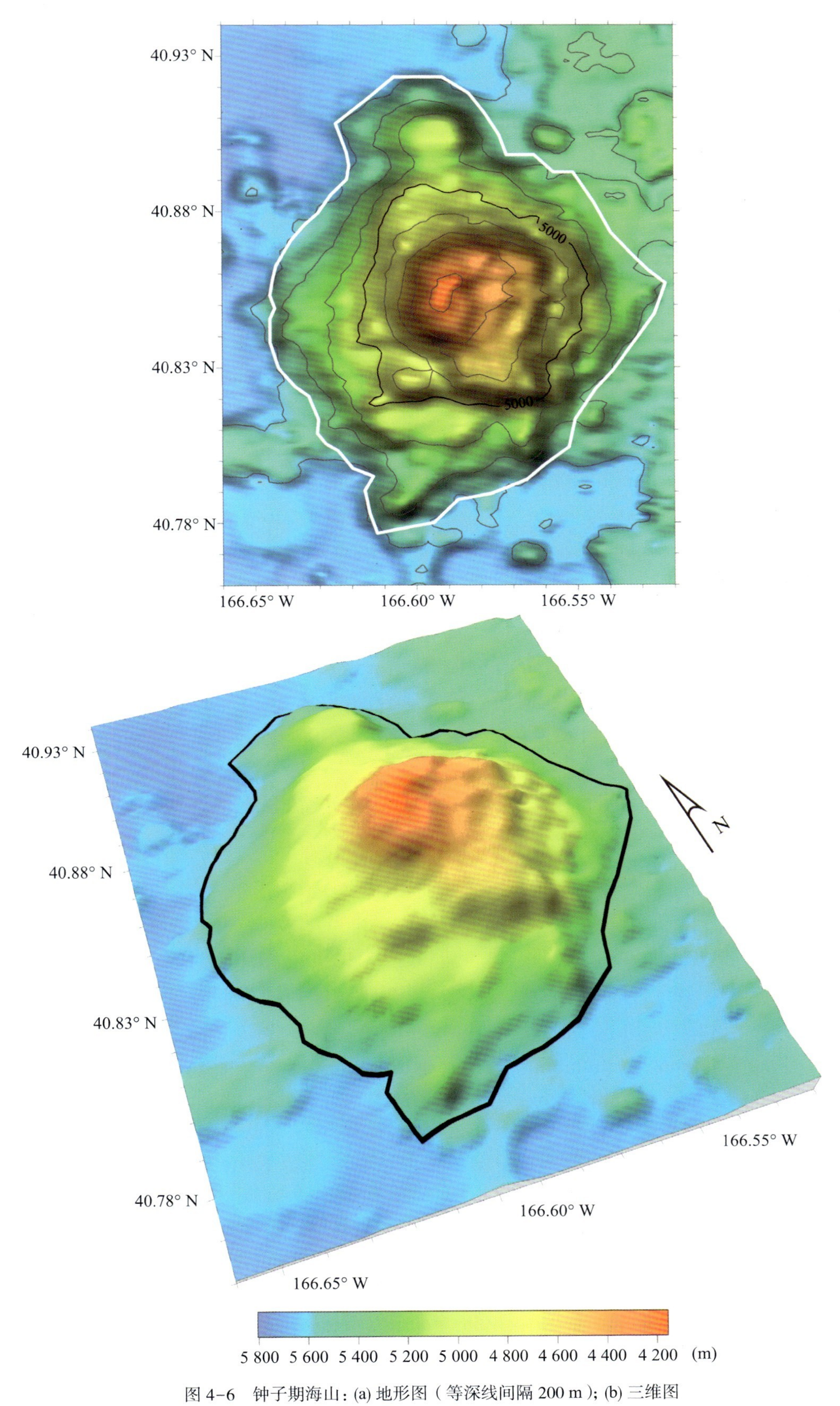

图 4-6 钟子期海山：(a) 地形图（等深线间隔 200 m）；(b) 三维图

Fig.4-6 Zhongziqi Seamount: (a) Bathymetric map (Contours are in 200 m); (b) 3-D topographic map

4.2.6 伯牙海山 Boya Seamount

<table>
<tr><td>中文名称
Chinese Name</td><td colspan="3">Bóyá Hǎishān
伯牙海山</td></tr>
<tr><td>英文名称
English Name</td><td colspan="3">Boya Seamount</td></tr>
<tr><td>地理区域 / Location</td><td colspan="3">东北太平洋 The Northeast Pacific Ocean</td></tr>
<tr><td rowspan="2">特征点坐标
Coordinate</td><td rowspan="2">40°32.7′N 166°40.3′W</td><td>长度 / Length</td><td>20 km</td></tr>
<tr><td>宽度 / Width</td><td>10 km</td></tr>
<tr><td>水深 / Depth</td><td>4 284~5 752 m</td><td>高差 / Total Relief</td><td>1 468 m</td></tr>
<tr><td>发现情况
Discovery Facts</td><td colspan="3">此海山于 2019 年“向阳红 06”船在执行航次调查时发现。
The seamount was discovered in 2019 during the survey cruise carried out onboard the Chinese R/V Xiang Yang Hong 06.</td></tr>
<tr><td>地形特征
Feature Description</td><td colspan="3">伯牙海山发育于太平洋东北部，音乐家海山群西北方向 1 111 km，呈 NE-SW 向延伸，长和宽分别为 20 km 和 10 km。该海山最高峰位于海山中部，峰顶水深 4 284 m，山麓水深 5 752 m，最大高差约 1 468 m。
Located in the Northeast Pacific Ocean, the seamount is 1,111 km northwest of the Musicians Seamounts. Extending in the NE-SW direction, it is 20 km long and 10 km wide. The highest peak of the seamount is located in the middle. It is 4,284 m deep at the top and 5,752 m deep at the foot, with a maximum total relief of about 1,468 m.</td></tr>
<tr><td>专名释义
Reason for Choice of Name</td><td colspan="3">该区域沿用音乐家相关命名体系群组化方法命名。伯牙，姓伯，名牙，春秋战国时期楚国郢都人，精通琴艺，其代表作有《高山流水》。《吕氏春秋》和《列子》中记载了伯牙与钟子期的故事，伯牙回家探亲时，在汉江边鼓琴，钟子期正巧遇见，感叹说“巍巍乎若高山，荡荡乎若流水”。因兴趣相投，两人成为至交。钟子期死后，伯牙认为世上再无知己，终生不再鼓琴。该海山位于钟子期海山南面，犹如伯牙席地而坐为钟子期鼓琴。故以伯牙命名。
The features in this sea area are all named following grouping naming system related to musicians. There was a famous musician named Boya, a native of Yingdu in the State of Chu during the Spring and Autumn Period and the Warring States Period of China. Being very good at playing stringed instruments, he had a representative work Lofty Mountains and Flowing Water. Mister Lv's Spring and Autumn Annals and Liezi recorded Boya and Zhong Ziqi's story in which Zhong Ziqi happened to meet Boya who was playing the Guqin at the waterfront of Han River when the latter returned home to visit relatives, and expressed his admiration of “as loft as the towering mountain, and as turbulent as the rushing water” . Because of similar interests, the two became very intimate friends. After Zhong Ziqi's death, Boya thought that there was no confidant in the world, hence never played the Guqin again from then on. The feature, which is to the south of Zhongziqi Seamount, looks like Boya sitting on the ground to play the Guqin for Zhong Ziqi. Therefore, the feature is named Boya Seamount after him.</td></tr>
</table>

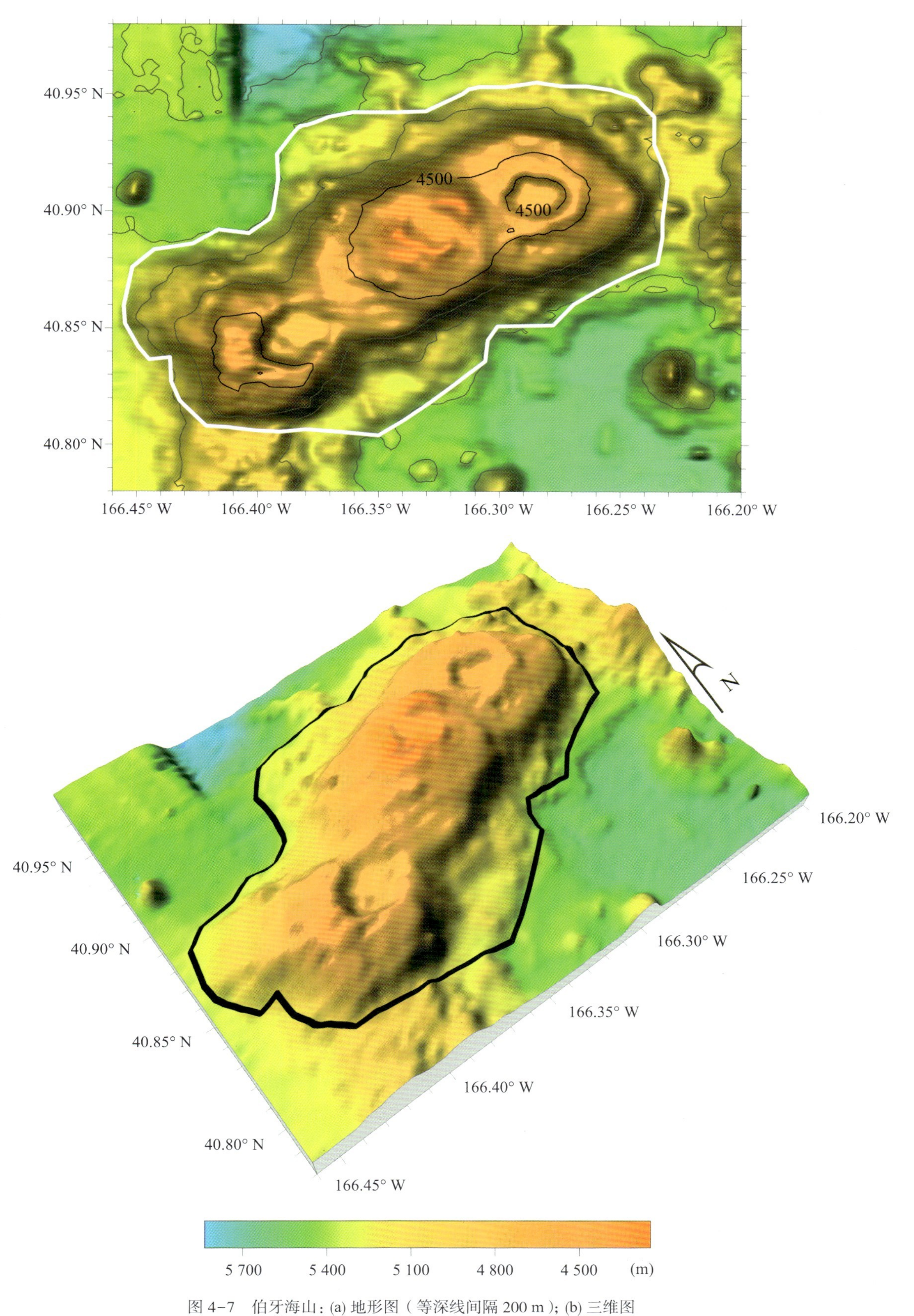

图 4-7 伯牙海山：(a) 地形图（等深线间隔 200 m）；(b) 三维图

Fig.4-7 Boya Seamount: (a) Bathymetric map (Contours are in 300 m); (b) 3-D topographic

4.2.7 桓谭海山 Huantan Seamount

<table>
<tr><td>中文名称
Chinese Name</td><td colspan="4">Huántán Hǎishān
桓谭海山</td></tr>
<tr><td>英文名称
English Name</td><td colspan="4">Huantan Seamount</td></tr>
<tr><td>地理区域 / Location</td><td colspan="4">东北太平洋 The Northeast Pacific Ocean</td></tr>
<tr><td rowspan="2">特征点坐标
Coordinate</td><td rowspan="2">40°53.4′N 166°20.5′W</td><td>长度 / Length</td><td colspan="2">11 km</td></tr>
<tr><td>宽度 / Width</td><td colspan="2">20 km</td></tr>
<tr><td>水深 / Depth</td><td>4 245~5 449 m</td><td>高差 / Total Relief</td><td colspan="2">1 204 m</td></tr>
<tr><td>发现情况
Discovery Facts</td><td colspan="4">此海山于 2019 年“向阳红 06”船在执行航次调查时发现。
The seamount was discovered in 2019 during the survey cruise carried out onboard the Chinese R/V Xiang Yang Hong 06.</td></tr>
<tr><td>地形特征
Feature Description</td><td colspan="4">桓谭海山位于东北太平洋，音乐家海山群以北 1 038 km，呈等维展布，南北长约 11 km，东西宽约 20 km。该海山主体为一圆柱状山峰，峰顶地形较为平坦，最高处位于海山西南侧，水深约为 4 245 m。
Located in the Northeast Pacific Ocean, the seamount is 1,038 km north of the Musicians Seamounts. Being equidimensional in plan, it is about 11 km long from south to north and about 20 km wide from west to east. The main body of the seamount is a cylindrical peak which is relatively flat at the top. The highest point is located in the southwest of the flat top, with a depth of about 4,245 m.</td></tr>
<tr><td>专名释义
Reason for
Choice of Name</td><td colspan="4">该区域沿用音乐家相关命名体系群组化方法命名。桓谭（约公元前 23—56），字君山，东汉哲学家、经学家、琴师、天文学家。爱好音律，善鼓琴，博学多通，遍习五经，能文章，尤其喜欢古学，还喜欢歌舞杂戏，为两汉之际著名学者。故以桓谭命名。
The features in this area are named following grouping naming system related to musicians. There was a famous musician named Huan Tan (approximately 23 BC-56 AD), with the courtesy name Junshan. He was a philosopher, scholar of Confucian classics, instrument master, and astronomer of the Eastern Han Dynasty. He liked music and was good at playing stringed instruments. Being versatile, he was familiar with the Five Classics and took a particular interest in ancient study, and was able to write high quality articles. Besides, he was fond of singing, dancing and dramas. He was a famous scholar between the Western Han Dynasty and the Eastern Han Dynasty. So the feature is named Huantan Seamount after him.</td></tr>
</table>

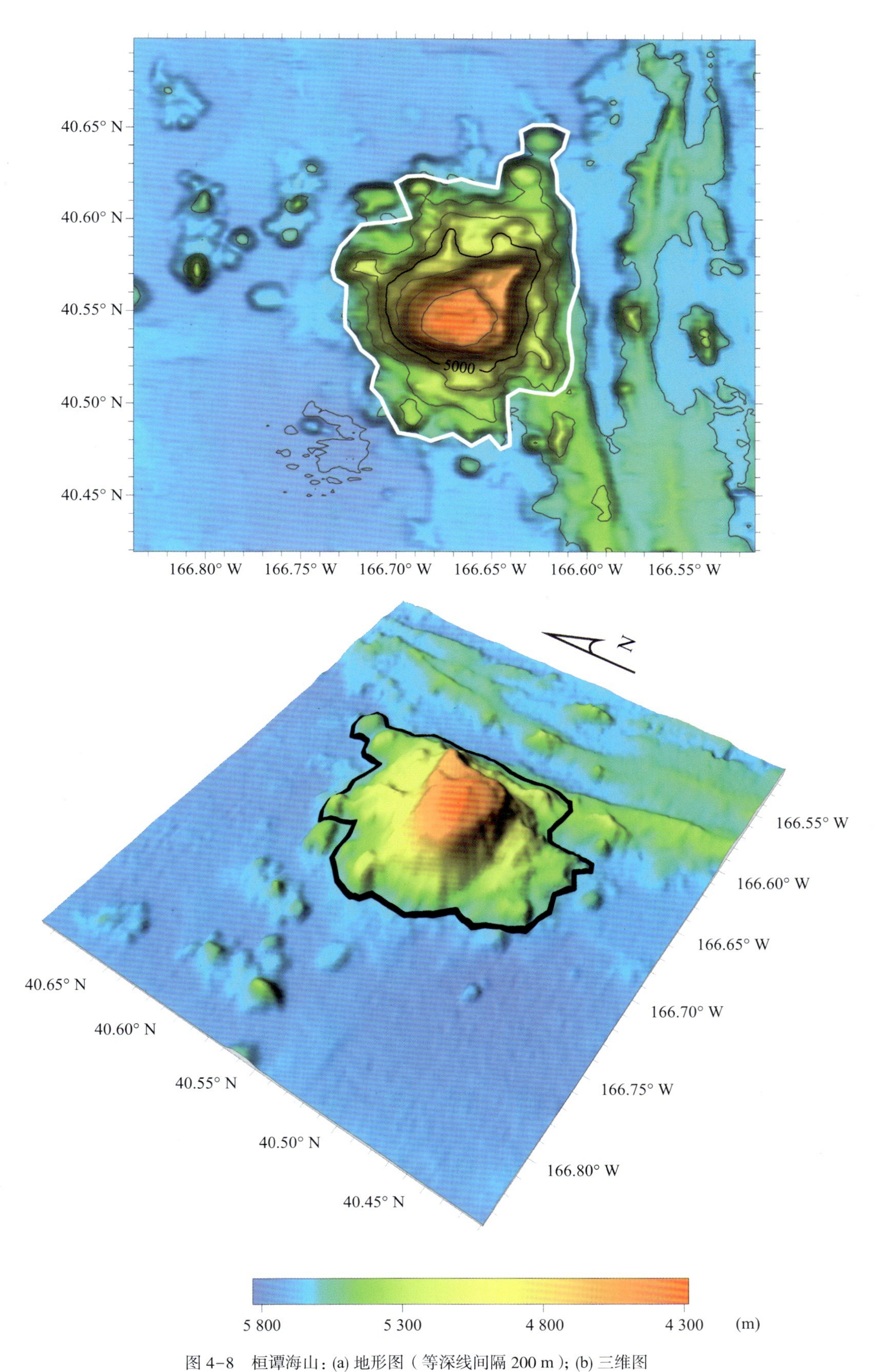

图 4-8 桓谭海山：(a) 地形图（等深线间隔 200 m）；(b) 三维图

Fig.4-8 Huantan Seamount: (a) Bathymetric map (Contours are in 200 m); (b) 3-D topographic map

4.2.8 蔡邕海山 Caiyong Seamount

<table>
<tr><td>中文名称
Chinese Name</td><td colspan="3">Caiyōng Hǎishān
蔡邕海山</td></tr>
<tr><td>英文名称
English Name</td><td colspan="3">Caiyong Seamount</td></tr>
<tr><td>地理区域 / Location</td><td colspan="3">东北太平洋 The Northeast Pacific Ocean</td></tr>
<tr><td rowspan="2">特征点坐标
Coordinate</td><td rowspan="2">39°22.0′N 165°36.5′W</td><td>长度 / Length</td><td>21 km</td></tr>
<tr><td>宽度 / Width</td><td>11 km</td></tr>
<tr><td>水深 / Depth</td><td>4 144~5 466 m</td><td>高差 / Total Relief</td><td>1 322 m</td></tr>
<tr><td>发现情况
Discovery Facts</td><td colspan="3">此海山于 2019 年“向阳红 06”船在执行航次调查时发现。
The seamount was discovered in 2019 during the survey cruise carried out onboard the Chinese R/V Xiang Yang Hong 06.</td></tr>
<tr><td>地形特征
Feature Description</td><td colspan="3">蔡邕海山发育于太平洋东北部，音乐家海山群以北 850 km，呈 E-W 向延伸，长和宽分别为 21 km 和 11 km，峰顶平缓，最高处位于海山东南部，水深 4 144 m，山麓水深 5 466 m，最大高差 1 322 m，西坡地形较陡，东坡地形平缓。
Located in the Northeast of the Pacific Ocean, the seamount is 850 km north of the Musicians Seamounts. Extending in the E-W direction, it is 21 km long and 11 km wide. The peak is flat, and the highest point, which is located in the southeast of the seamount, is 4,144 m deep, whereas the piedmont of the seamount is 5,466 m deep, with a maximum total relief of 1,322 m. It is relatively steep on the western slope and relatively gentle on the eastern slope.</td></tr>
<tr><td>专名释义
Reason for Choice of Name</td><td colspan="3">该区域沿用音乐家相关命名体系群组化方法命名。蔡邕(133—192)，字伯喈。东汉时期名臣，文学家、书法家，才女蔡文姬之父，精通音律，才华横溢，师事著名学者胡广。除通经史、善辞赋之外，还精于书法，所创“飞白”书体，对后世影响甚大。故以蔡邕命名。
The features in this area are all named following grouping naming system related to musicians. There was a famous musician named Cai Yong (133AD-192AD), with the courtesy name Bojie. He was a famous official, writer and calligrapher of the Eastern Han Dynasty. He was also a master of music, and the father of Cai Wenji, a talented woman. By learning from the famous scholar Hu Guang, he was familiar with classics and historical books, and was good at classical Chinese writing. Moreover, he was adept at calligraphy, and created Feibai Calligraphy, which exerted great impact on later generations. So the feature is named Caiyong Seamount after him.</td></tr>
</table>

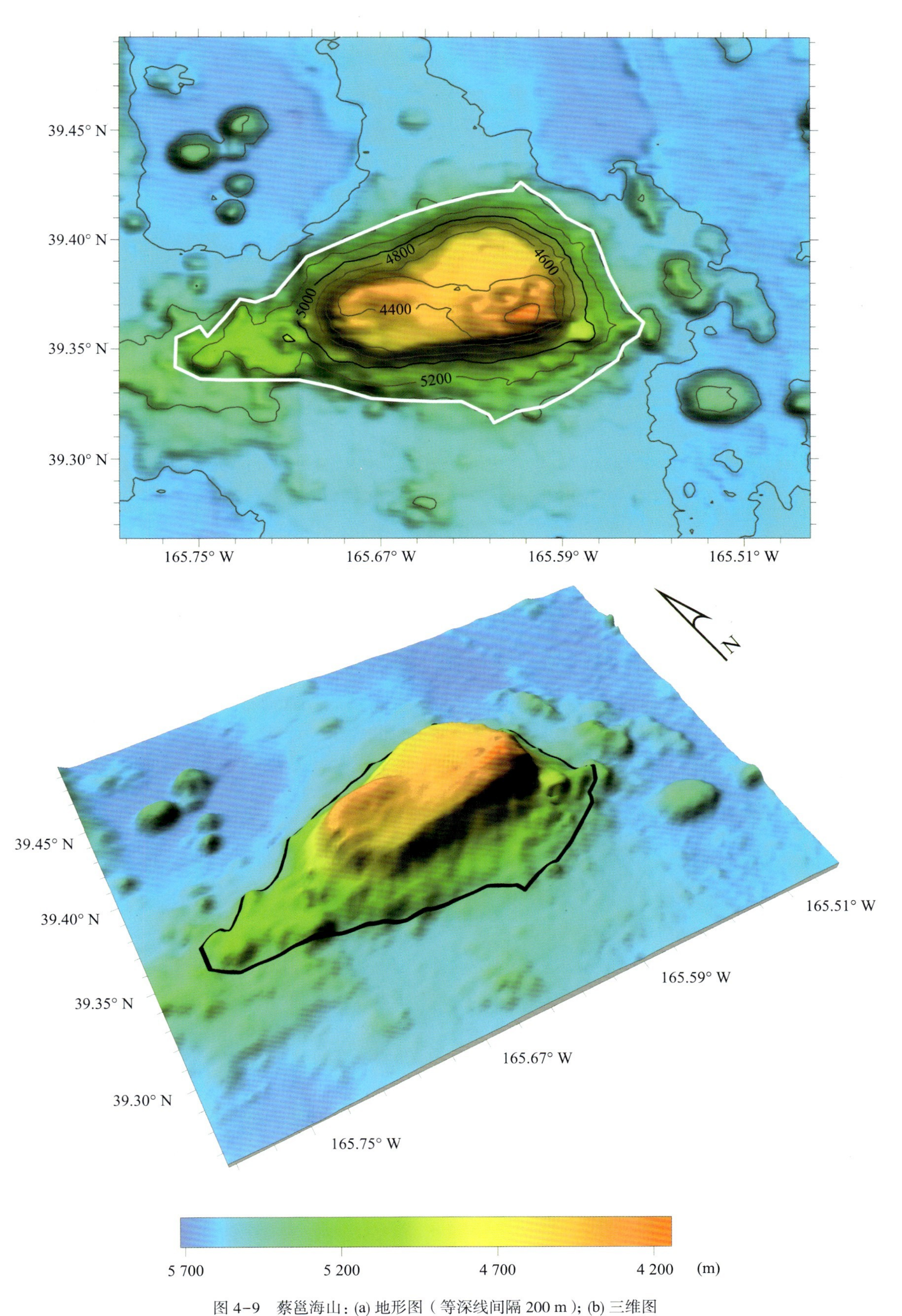

图 4-9 蔡邕海山：(a) 地形图（等深线间隔 200 m）；(b) 三维图

Fig.4-9 Caiyong Seamount: (a) Bathymetric map (Contours are in 200 m); (b) 3-D topographic map

4.2.9 万宝常海山群 Wanbaochang Seamounts

<table>
<tr><td>中文名称
Chinese Name</td><td colspan="3">Wànbǎocháng Hǎishānqún
万宝常海山群</td></tr>
<tr><td>英文名称
English Name</td><td colspan="3">Wanbaochang Seamounts</td></tr>
<tr><td>地理区域 / Location</td><td colspan="3">东北太平洋 The Northeast Pacific Ocean</td></tr>
<tr><td rowspan="2">特征点坐标
Coordinate</td><td rowspan="2">39°56.8′N 162°40.6′W
39°53.2′N 162°42.4′W
39°48.9′N 162°48.7′W</td><td>长度 / Length</td><td>48 km</td></tr>
<tr><td>宽度 / Width</td><td>14 km</td></tr>
<tr><td>水深 / Depth</td><td>4 450~5 854 m</td><td>高差 / Total Relief</td><td>1 404 m</td></tr>
<tr><td>发现情况
Discovery Facts</td><td colspan="3">此海山群于 2020 年“向阳红 01”船在执行航次调查时发现。
The seamounts were discovered in 2020 during the survey cruise carried out onboard the Chinese R/V Xiang Yang Hong 01.</td></tr>
<tr><td>地形特征
Feature Description</td><td colspan="3">万宝常海山群发育于太平洋东北部，位于音乐家海山群以北 915 km，由 6 座圆形海山组成，整体呈 NE-SW 向条状延伸，长和宽分别为 48 km 和 14 km。其最高峰位于海山群东北部圆顶，水深 4 450 m，山麓水深 5 854 m，最大高差约 1 404 m。
Located in the northeast of the Pacific Ocean, the seamounts are 915 km north of the Musicians Seamounts. Consisting of 6 rounded seamounts, they extend in the NE-SW direction as stripe as whole, and are 48 km long and 14 km wide. The highest point is located at the dome in the northeast of the seamounts, with a depth of 4,450 m. The piedmont of the seamounts is 5,854 m deep. Hence, the maximum total relief is about 1,404 m.</td></tr>
<tr><td>专名释义
Reason for Choice of Name</td><td colspan="3">该区域沿用音乐家相关命名体系群组化方法命名。万宝常，隋代音乐家，幼学琵琶，师事北齐中书侍郎祖珽，精通多种乐器，深知乐理和声律。故以万宝常命名。
The features in this area are all named following grouping naming system related to musicians. There was a famous musician named Wan Baochang, a musician of the Sui Dynasty. He learned to play the Pipa at a young age, and was a student of Zu Ting, an official of the Department of Imperial Secretariat of the Northern Qi Dynasty. He was proficient in many musical instruments and had a solid understanding of music theories and rhythms. So the features are named Wanbaochang Seamounts after him.</td></tr>
</table>

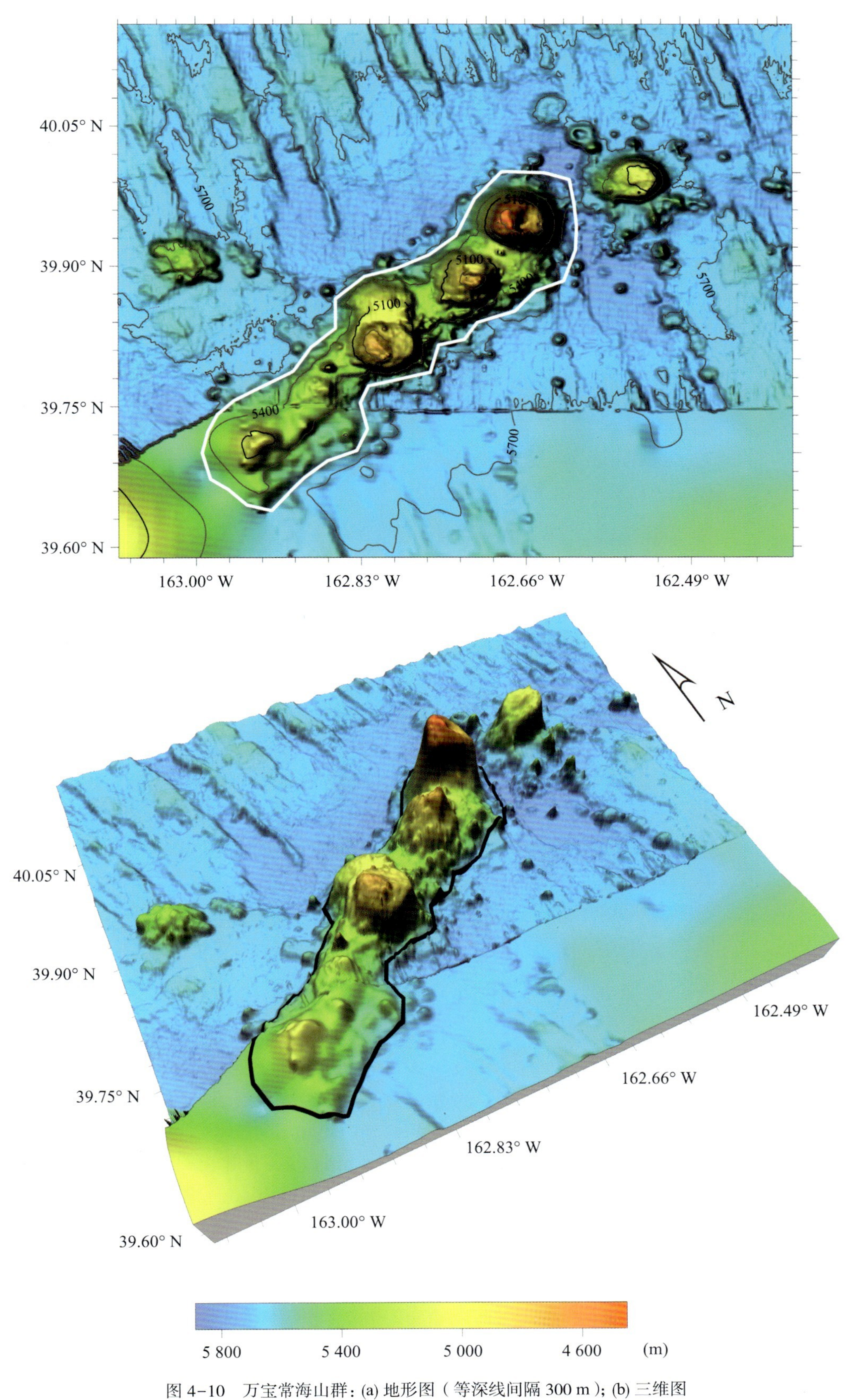

图 4-10 万宝常海山群：(a) 地形图（等深线间隔 300 m）；(b) 三维图

Fig.4-10 Wanbaochang Seamounts: (a) Bathymetric map (Contours are in 300 m); (b) 3-D topographic map

4.2.10 苏祇婆海山 Suzhipo Seamount

中文名称 Chinese Name	Sūzhīpó Hǎishān 苏祇婆海山		
英文名称 English Name	Suzhipo Seamount		
地理区域 / Location	东北太平洋 The Northeast Pacific Ocean		
特征点坐标 Coordinate	40°53.6′N 162°17.2′W	**长度 / Length**	23 km
		宽度 / Width	15 km
水深 / Depth	4 425~5 638 m	**高差 / Total Relief**	1 213 m
发现情况 Discovery Facts	此海山于 2020 年“向阳红 01”船在执行航次调查时发现。 The seamount was discovered in 2020 during the survey cruise carried out onboard the Chines R/V *Xiang Yang Hong 01.*		
地形特征 Feature Description	苏祇婆海山发育于太平洋东北部，位于音乐家海山群以北 1 080 km，整体呈等维展布，立体呈锥状，长和宽分别为 23 km 和 15 km。该海山顶端突出，峰顶水深 4 425 m，山麓水深 5 638 m，最大高差约 1 213 m。该海山孤立于深海平原中，周围密布 NW-SE 走向小型沟槽，周边平均水深超过 5 200 m。 Located in the northeast of the Pacific Ocean, the seamount is 1,080 km north of the Musicians Seamounts. Being equidimensional in plan as a whole, it takes the shape of a 3-dimentional cone, and is 23 km long and 15 km wide. The seamount protrudes from the top, and it is 4,425 m deep at the summit, and 5,638 m at the piedmont, with a maximum total relief of about 1,213 m. Being isolated on the abyssal plain, the seamount is densely covered by many small trenches extending in the NW-SE direction, with an average depth of more than 5,200 m in the surroundings.		
专名释义 Reason for Choice of Name	该区域沿用音乐家相关命名体系群组化方法命名。苏祇婆，西域龟兹（现中国新疆维吾尔自治区）人，北周至隋代著名的音乐家、琵琶演奏家。故以苏祇婆命名。 The features in this sea area are all named following grouping naming system related to musicians. There was a famous musician named Su Zhipo, a native of Qiuci, an ancient state in the Western Regions (now Xinjiang Uygur Autonomous Region of China). He was a famous musician and Pipa player from the Northern Zhou Dynasty to the Sui Dynasty. Hence, the feature is named Suzhipo Seamount after him.		

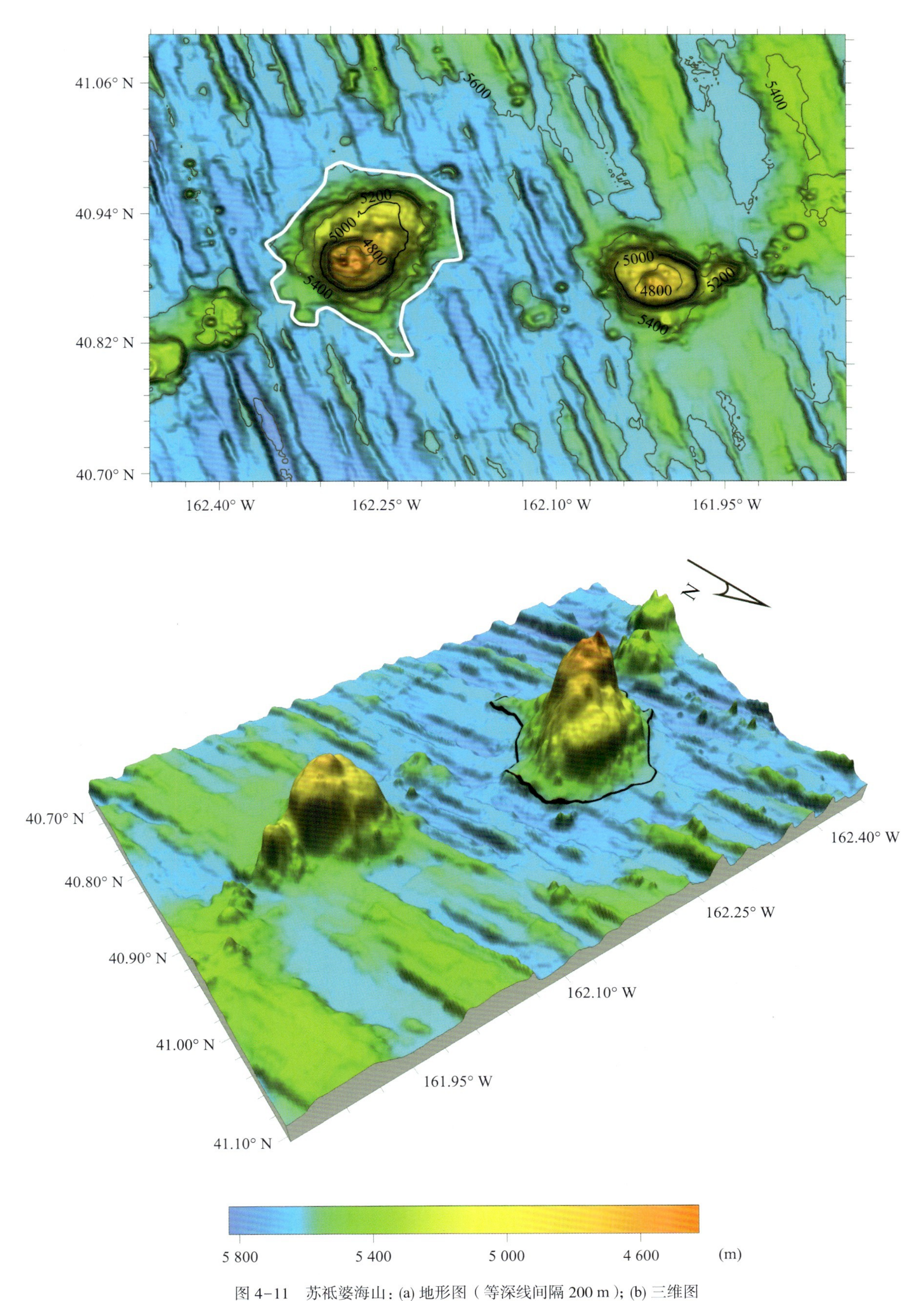

图 4-11 苏祗婆海山：(a) 地形图（等深线间隔 200 m）；(b) 三维图

Fig.4-11 Suzhipo Seamount: (a) Bathymetric map (Contours are in 200 m); (b) 3-D topographic map

4.2.11 李龟年海山 Liguinian Seamount

<table>
<tr><td>中文名称
Chinese Name</td><td colspan="3">Lǐguīnián Hǎishān
李龟年海山</td></tr>
<tr><td>英文名称
English Name</td><td colspan="3">Liguinian Seamount</td></tr>
<tr><td>地理区域 / Location</td><td colspan="3">东北太平洋 The Northeast Pacific Ocean</td></tr>
<tr><td rowspan="2">特征点坐标
Coordinate</td><td rowspan="2">40°40.9′N 160°20.0′W</td><td>长度 / Length</td><td>37 km</td></tr>
<tr><td>宽度 / Width</td><td>17 km</td></tr>
<tr><td>水深 / Depth</td><td>2 845~5 454 m</td><td>高差 / Total Relief</td><td>2 609 m</td></tr>
<tr><td>发现情况
Discovery Facts</td><td colspan="3">此海山于 2019 年“向阳红 06”船在执行航次调查时发现。
The seamount was discovered in 2019 during the survey cruise carried out onboard the Chinese R/V Xiang Yang Hong 06.</td></tr>
<tr><td>地形特征
Feature Description</td><td colspan="3">李龟年海山发育于太平洋东北部，位于音乐家海山群以北 1 113 km，整体呈 NE-SW 向条状延伸，长和宽分别为 37 km 和 17 km。该海山发育 2 个规模相似的小型海山，最高处位于西南部峰顶，水深 2 845 m，底部水深 5454m。该海山孤立于深海平原中，周边广泛分布小型凸起。
Located in the northeast of the Pacific Ocean, the seamount is 1,113 km north of the Musicians Seamounts. Extending in the NE-SW direction as stripe as whole, it is 37 km long and 17 km wide. There are two small seamounts of similar sizes in the seamount. The highest point, the summit in the southwest, is 2,845 m deep whereas the foot of the mountain is 5,454 m deep. Isolated on the abyssal plain, the seamount is surrounded by small protrusions extensively.</td></tr>
<tr><td>专名释义
Reason for Choice of Name</td><td colspan="3">该区域沿用音乐家相关命名体系群组化方法命名。李龟年，唐代音乐家，被后人誉为“唐代乐圣”，精通音律，通晓多种演奏技巧，尤善吹奏笛子和筚篥。故以李龟年命名。
The features in this sea area are all named following grouping naming system related to musicians. There was a famous musician named Li Guinian, a musician of the Tang Dynasty. Hailed as the Music Sage of the Tang Dynasty by later generations, he was a master of music and possessed a variety of performance skills, and played the flute and the Bili quite well. So the feature is named Liguinian Seamount after him.</td></tr>
</table>

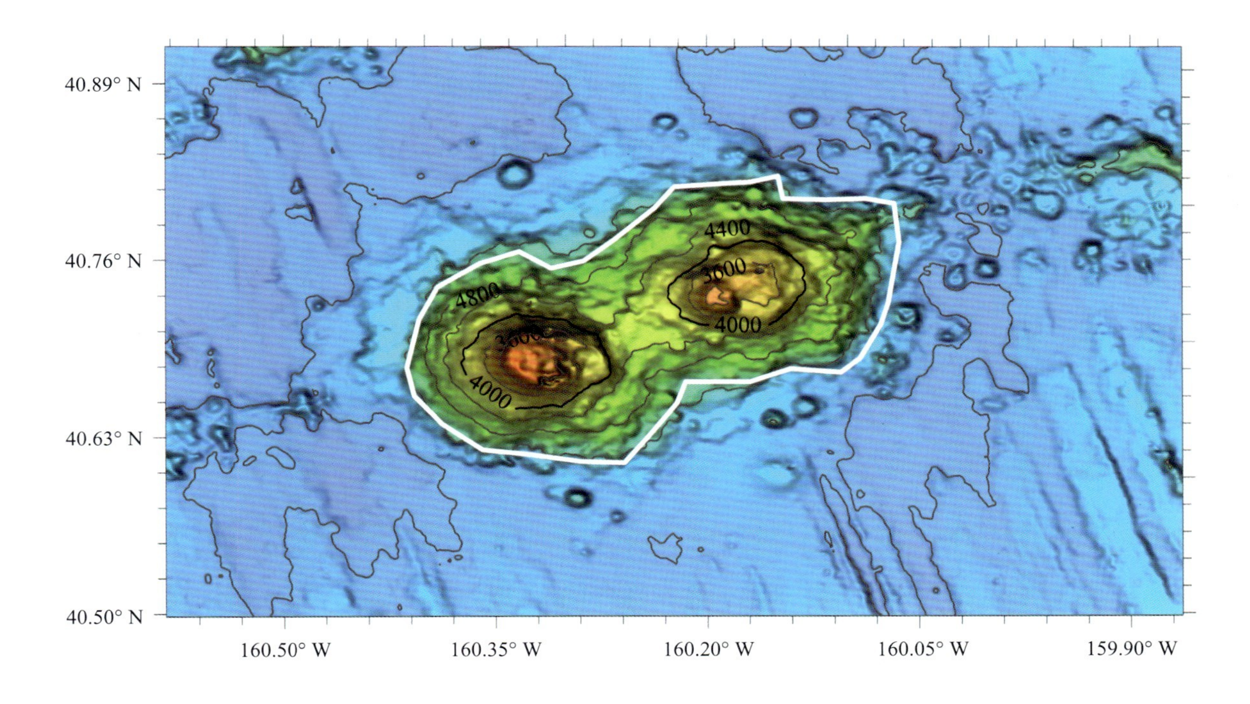

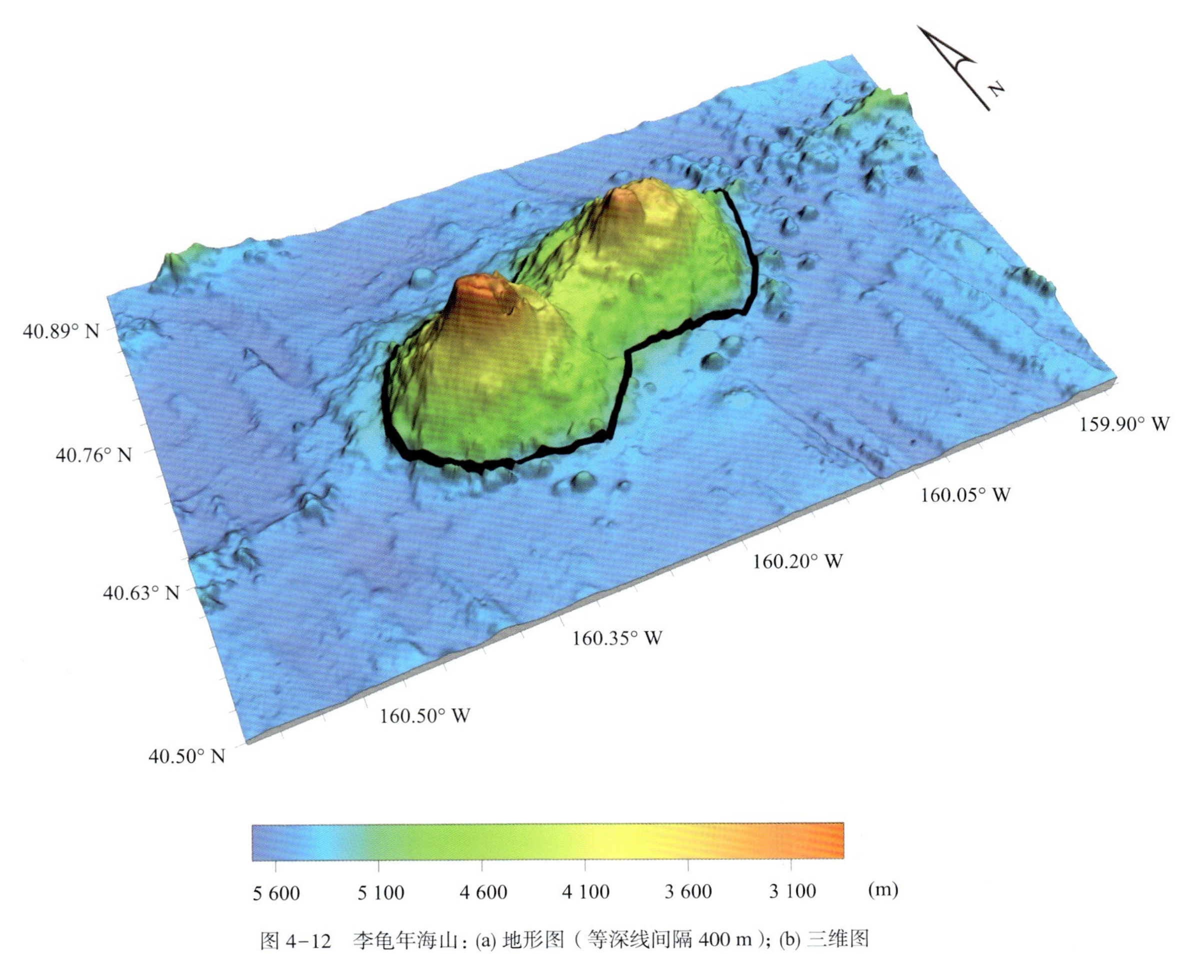

图 4-12 李龟年海山：(a) 地形图（等深线间隔 400 m）；(b) 三维图

Fig.4-12 Liguinian Seamount: (a) Bathymetric map (Contours are in 400 m); (b) 3-D topographic map

4.2.12 段善本海山 Duanshanben Seamount

<table>
<tr><td>中文名称
Chinese Name</td><td colspan="4">Duànshànběn Hǎishān
段善本海山</td></tr>
<tr><td>英文名称
English Name</td><td colspan="4">Duanshanben Seamount</td></tr>
<tr><td>地理区域 / Location</td><td colspan="4">东北太平洋 The Northeast Pacific Ocean</td></tr>
<tr><td rowspan="2">特征点坐标
Coordinate</td><td rowspan="2">39°28.4′N 161°05.1′W</td><td>长度 / Length</td><td colspan="2">17 km</td></tr>
<tr><td>宽度 / Width</td><td colspan="2">12 km</td></tr>
<tr><td>水深 / Depth</td><td>4 591~5 710 m</td><td>高差 / Total Relief</td><td colspan="2">1 119 m</td></tr>
<tr><td>发现情况
Discovery Facts</td><td colspan="4">此海山于 2019 年“向阳红 06”船在执行航次调查时发现。
The seamount was discovered in 2019 during the survey cruise carried out onboard the Chinese R/V Xiang Yang Hong 06.</td></tr>
<tr><td>地形特征
Feature Description</td><td colspan="4">段善本海山发育于太平洋东北部，位于音乐家海山群以北 919 km，呈 NE-SW 向延伸，长和宽分别为 17 km 和 12 km。该海山发育 2 个小型海山，最高处位于东北部较大海山峰顶，水深 4 591 m，山麓水深 5 710 m。该海山孤立于深海平原中，周边广泛分布小型沟槽和凸起。
Located in the northeast of the Pacific Ocean, the seamount is 919 km north of the Musicians Seamounts. Extending in the NE-SW direction, it is 17 km long and 12 km wide. The seamount consists of two small seamounts. The highest point, which is the summit of the larger seamount in the northeast, is 4,591 m deep, whereas the piedmont of the mountain is 5,710 m deep. Isolated on the abyssal plain, the seamount is surrounded by small trenches and protrusions.</td></tr>
<tr><td>专名释义
Reason for Choice of Name</td><td colspan="4">该区域沿用音乐家相关命名体系群组化方法命名。段善本，唐代琵琶演奏家，长安庄严寺僧，法名善本。琵琶技艺高超，人称“段师”，作有《西梁州》《道调凉州》等曲。故以段善本命名。
The features in this sea area are all named following grouping naming system related to musicians. There was a famous musician named Duan Shanben, a Pipa expert of the Tang Dynasty. He was also a monk of the Zhuangyan Temple of Chang′an (Xi′an now), with a religious name of Shanben. With superb Pipa playing skills, he was called Master Duan. His representative works included Xiliangzhou and Daodiao Liangzhou. So the feature is named Duanshanben Seamount after him.</td></tr>
</table>

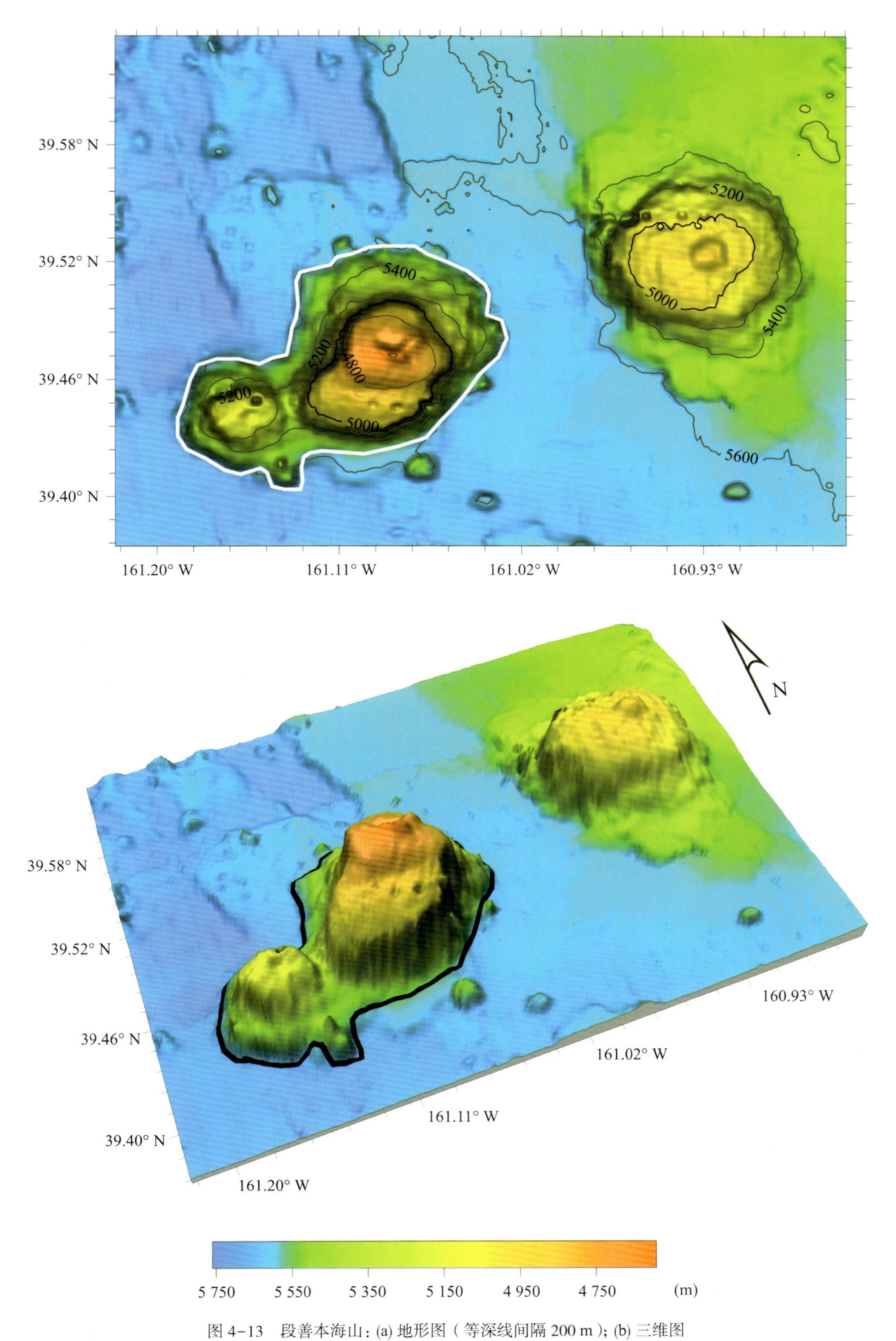

图 4-13 段善本海山：(a) 地形图（等深线间隔 200 m）；(b) 三维图

Fig.4-13 Duanshanben Seamount: (a) Bathymetric map (Contours are in 200 m); (b) 3-D topographic map

4.2.13 姜夔海山群 Jiangkui Seamounts

<table>
<tr><td>中文名称
Chinese Name</td><td colspan="3">Jiāngkuí Hǎishānqún
姜夔海山群</td></tr>
<tr><td>英文名称
English Name</td><td colspan="3">Jiangkui Seamounts</td></tr>
<tr><td>地理区域 / Location</td><td colspan="3">东北太平洋 The Northeast Pacific Ocean</td></tr>
<tr><td rowspan="2">特征点坐标
Coordinate</td><td rowspan="2">39°14.9′N 160°50.4′W
39°12.8′N 161°01.2′W
39°12.0′N 161°07.3′W</td><td>长度 / Length</td><td>46 km</td></tr>
<tr><td>宽度 / Width</td><td>25 km</td></tr>
<tr><td>水深 / Depth</td><td>3 524~5 720 m</td><td>高差 / Total Relief</td><td>2 196 m</td></tr>
<tr><td>发现情况
Discovery Facts</td><td colspan="3">此海山群于 2019 年“向阳红 06”船在执行航次调查时发现。
The seamounts were discovered in 2019 during the survey cruise carried out onboard the Chinese R/V Xiang Yang Hong 06.</td></tr>
<tr><td>地形特征
Feature Description</td><td colspan="3">姜夔海山群发育于太平洋东北部，位于音乐家海山群以北 881 km，呈 E-W 走向，长和宽分别为 46 km 和 25 km。该海山群由三个小型海山自东向西依次排列组成，最高处位于东部海山峰顶，水深 3 524 m。该海山孤立于深海平原中，周边水深 5 720 m。
Located in the northeast of the Pacific Ocean, the seamounts are 881 km north of the Musicians Seamounts. Extending in the E-W direction, they are 46 km long and 25 km wide. The seamounts consist of three smaller seamounts extending from east to west. The highest point, which is 3,524 m deep, is located at the summit of the eastern seamount. The seamounts are isolated on the abyssal plain, with a depth of 5,720 m in the surroundings.</td></tr>
<tr><td>专名释义
Reason for Choice of Name</td><td colspan="3">该区域沿用音乐家相关命名体系群组化方法命名。姜夔（约 1155—约 1221），字尧章，号白石道人，汉族，南宋文学家、音乐家，对诗词、散文、书法、音乐，无不精善。故以姜夔命名。
The features in this sea area are all named following grouping naming system related to musicians. There was a famous musician named Jiang Kui (approximately 1155AD–1221AD), with the courtesy name Yaozhang and literary name Baishidaoren. He was a writer and musician of the Southern Song Dynasty, skilled in poetry, prose, calligraphy and music. So the features are named Jiangkui Seamounts after him.</td></tr>
</table>

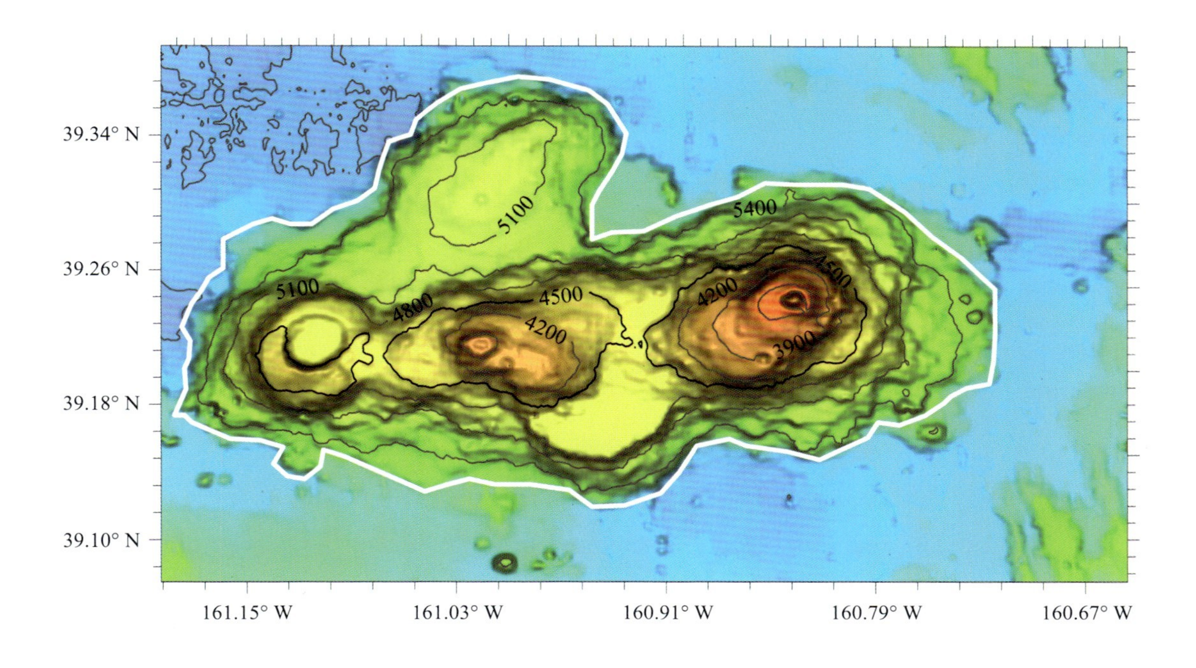

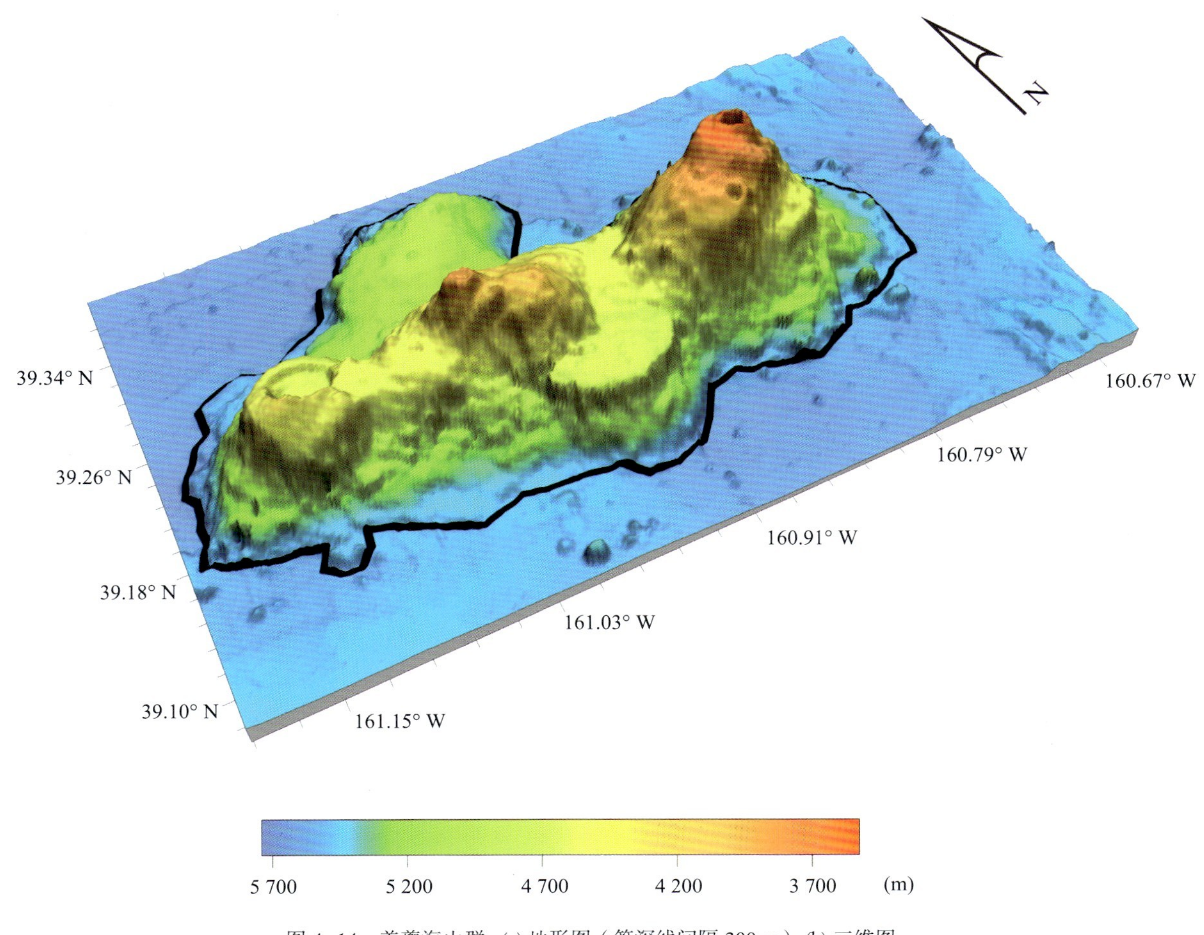

图 4-14 姜夔海山群：(a) 地形图（等深线间隔 300 m）；(b) 三维图

Fig.4-14 Jiangkui Seamounts: (a) Bathymetric map (Contours are in 300 m); (b) 3-D topographic map

4.2.14 雷海青海山群 Leihaiqing Seamounts

中文名称 Chinese Name	Léihǎiqīng Hǎishānqún 雷海青海山群		
英文名称 English Name	Leihaiqing Seamounts		
地理区域 / Location	东北太平洋 The Northeast Pacific Ocean		
特征点坐标 Coordinate	39°00.0′N 160°53.9′W 39°01.3′N 160°46.7′W	长度 / Length 宽度 / Width	24 km 9 km
水深 / Depth	4 401~5 615 m	高差 / Total Relief	1 214 m
发现情况 Discovery Facts	此海山群于 2019 年“向阳红 06”船在执行航次调查时发现。 The seamounts were discovered in 2019 during the survey cruise carried out onboard the Chinese R/V *Xiang Yang Hong 06.*		
地形特征 Feature Description	雷海青海山群发育于太平洋东北部，位于音乐家海山群以北 865 km，呈 E-W 带状走向，长和宽分别为 24 km 和 9 km。该海山群由 2 个规模相似的小型海山自东向西依次排列组成，俯视西部海山呈椭圆形，东部海山呈圆形，两个海山边缘陡峭。最高处位于西部海山，峰顶水深 4 401 m，山麓水深 5 615 m。该海山群孤立于深海平原中，周边水深 5 700 m。 Located in the northeast of the Pacific Ocean, the seamounts are 865 km north of the Musicians Seamounts. Extending in the E-W direction like a belt, they are 24 km long and 9 km wide. Leihaiqing Seamounts consist of two smaller seamount of similar size extending from east to west. The western seamount is oval and the eastern seamount is rounded when viewed from above, and edges of the two seamounts are steep. The highest point is located in the western seamount, which is 4,401 m deep at the summit, and 5,615 m deep at the piedmont. The seamounts are isolated on the abyssal plain, with a depth of 5,700 m in the surroundings.		
专名释义 Reason for Choice of Name	该区域沿用音乐家相关命名体系群组化方法命名。雷海青（696—756），是唐玄宗时的著名宫廷乐师，善弹琵琶，是福建、台湾及广东潮州一带艺人供奉的戏神，也是闽台民间信仰的重要神祇之一。故以雷海青命名。 The features in this sea area are all named following grouping naming system related to musicians. There was a famous musician named Lei Haiqing (696AD–756 AD). He was a famous court music master during the Reign of Emperor Xuanzong of the Tang Dynasty. Being good at playing the Pipa, he was a theater god enshrined by artists in Fujian, Taiwan, and Chaozhou of Guangdong Province, and one of the important gods in folk beliefs of Fujian and Taiwan. So the features are named Leihaiqing Seamounts after him.		

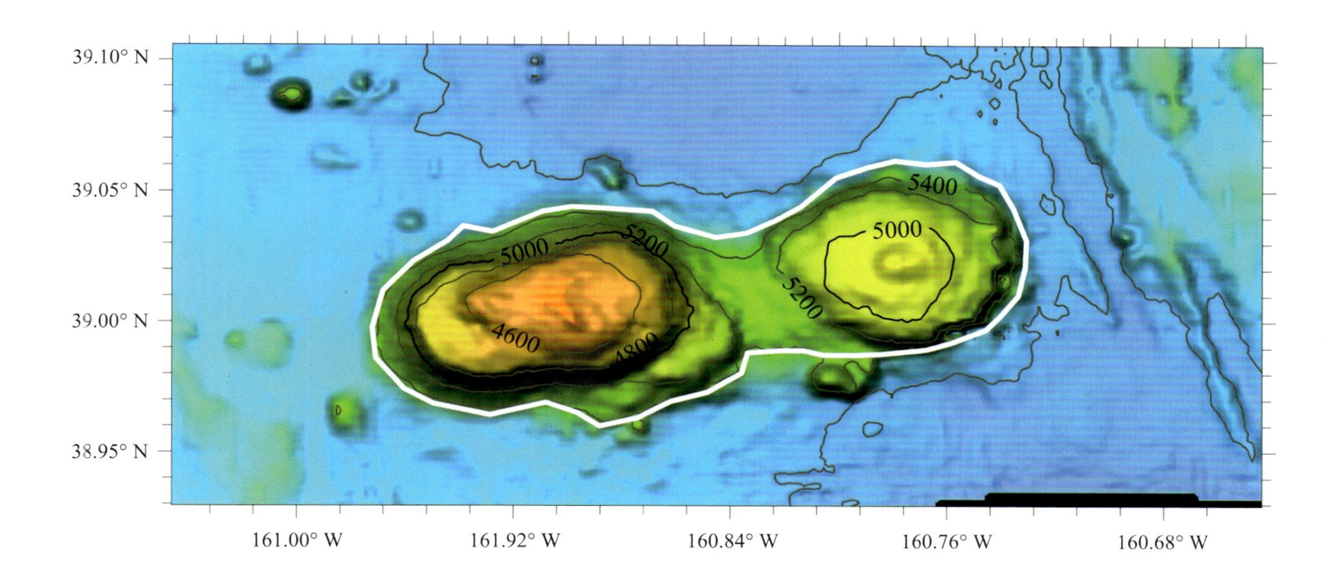

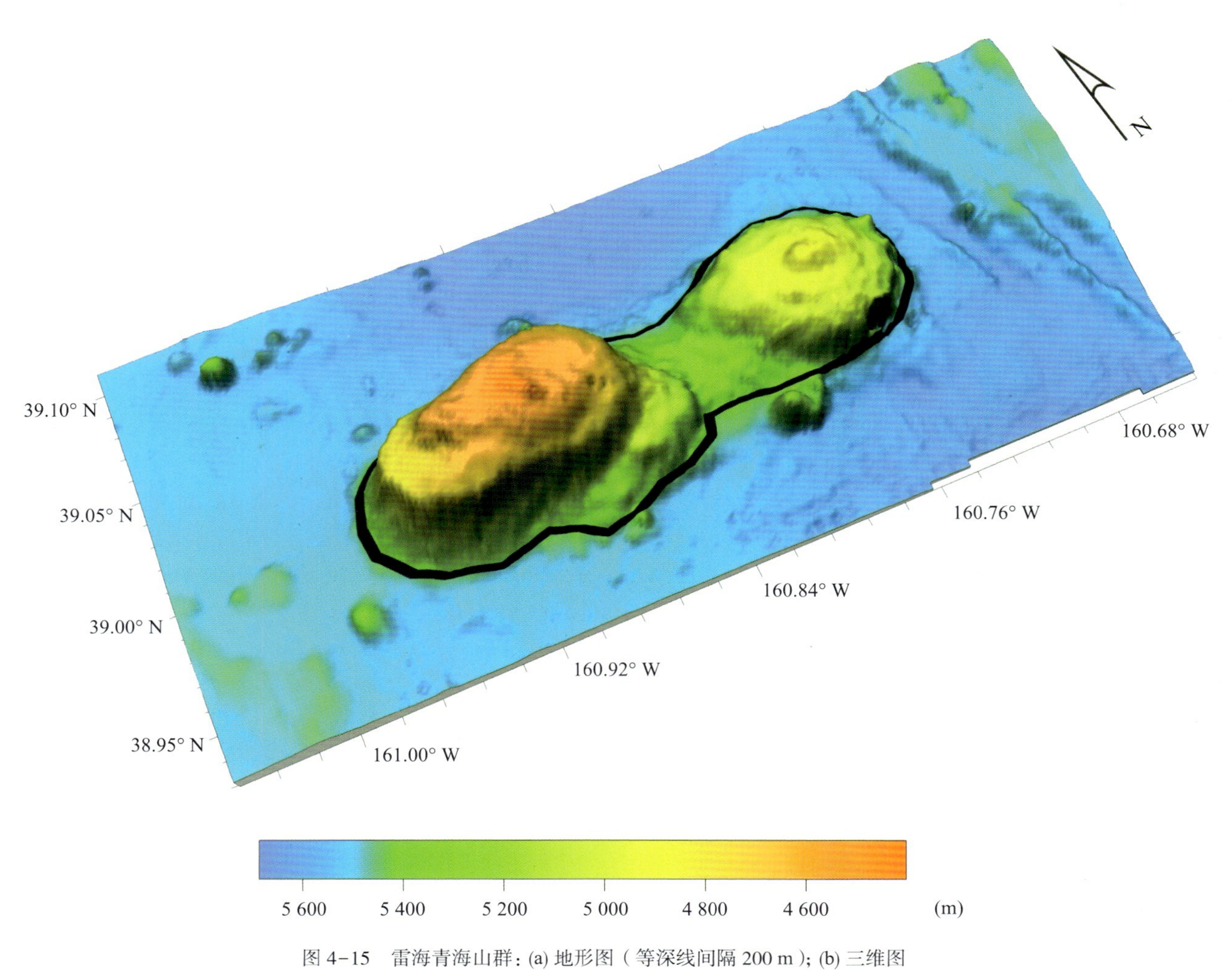

图 4-15 雷海青海山群：(a) 地形图（等深线间隔 200 m）；(b) 三维图

Fig.4-15 Leihaiqing Seamounts: (a) Bathymetric map (Contours are in 200 m); (b) 3-D topographic map

4.2.15 朱载堉海山 Zhuzaiyu Seamount

中文名称 Chinese Name	Zhūzǎiyù Hǎishān 朱载堉海山		
英文名称 English Name	Zhuzaiyu Seamount		
地理区域 / Location	东北太平洋 The Northeast Pacific Ocean		
特征点坐标 Coordinate	39°33.5′N 157°58.2′W	长度 / Length	15 km
		宽度 / Width	11 km
水深 / Depth	3 828~5 382 m	高差 / Total Relief	1 554 m
发现情况 Discovery Facts	此海山于2019年“向阳红06”船在执行航次调查时发现。 The seamount was discovered in 2019 during the survey cruise carried out onboard the Chinese R/V *Xiang Yang Hong 06*.		
地形特征 Feature Description	朱载堉海山发育于太平洋东北部，位于音乐家海山群东北方向1 100 km。整体呈圆形，长和宽分别为15 km和11 km。该海山中部突出，边缘陡峭，峰顶水深3 828 m，底部水深5 382 m，最大高差约1 554 m。 Located in the northeast of the Pacific Ocean, the seamount is 1,100 km northeast of the Musicians Seamounts. Being rounded in shape as a whole, it is 15 km long and 11 km wide. There are protrusions in the middle of the seamount, and its edges are steep. The peak is 3,828 m deep whereas the foot is 5,382 m deep, with a maximum total relief of about 1,554 m.		
专名释义 Reason for Choice of Name	该区域沿用音乐家相关命名体系群组化方法命名。朱载堉(1536—1611)，男，字伯勤，号句曲山人、九峰山人。明代著名的律学家（有“律圣”之称）、历学家、音乐家、创建了十二平均律。故以朱载堉命名。 The features in this area are all named following grouping naming system related to musicians. There was a famous musician named Zhu Zaiyu (1536AD–1611 AD), with the courtesy name Boqin and literature name Juqushanren. He was a famous legal scholar (known as the “Sage of Laws”), calendar scholar, and musician of the Ming Dynasty creating the twelve-tone equal temperament. So the feature is named Zhuzaiyu Seamount after him.		

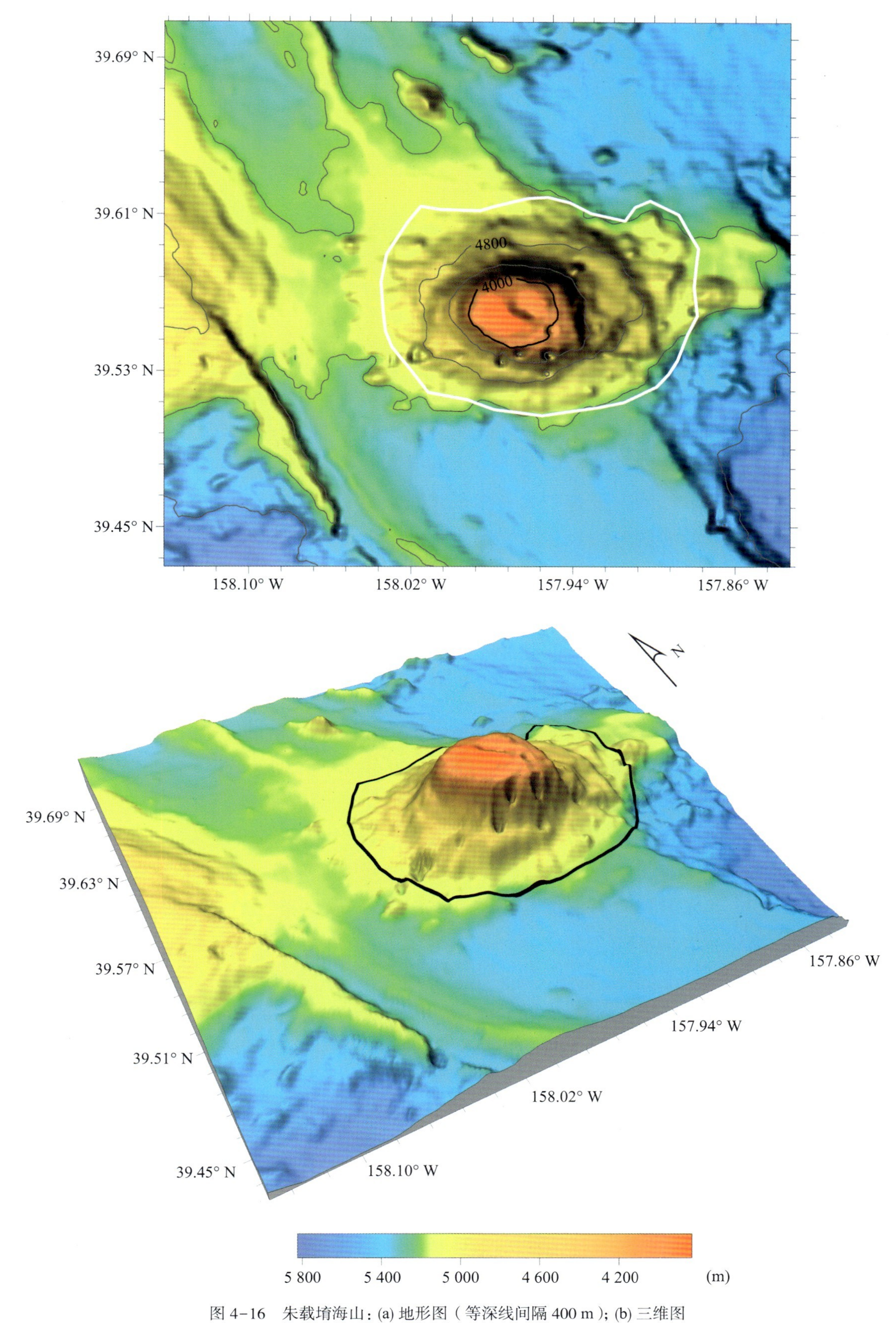

图 4-16 朱载堉海山：(a) 地形图（等深线间隔 400 m）；(b) 三维图

Fig.4-16 Zhuzaiyu Seamount: (a) Bathymetric map (Contours are in 400 m); (b) 3-D topographic map

4.2.16 魏良辅海山 Weiliangfu Seamount

中文名称 Chinese Name	Wèiliángfŭ Hăishān 魏良辅海山		
英文名称 English Name	Weiliangfu Seamount		
地理区域 / Location	东北太平洋 The Northeast Pacific Ocean		
特征点坐标 Coordinate	39°36.0′N 157°45.2′W	长度 / Length	12 km
		宽度 / Width	10 km
水深 / Depth	4 205~5 527 m	高差 / Total Relief	1 322 m
发现情况 Discovery Facts	此海山于 2019 年“向阳红 06”船在执行航次调查时发现。 The seamount was discovered in 2019 during the survey cruise carried out onboard of the Chinese R/V *Xiang Yang Hong 06*.		
地形特征 Feature Description	魏良辅海山发育于太平洋东北部，位于音乐家海山群东北方向 1 120 km，整体呈椭圆形，长和宽分别为 12 km 和 10 km。该海山中部地势西高东低，边缘陡峭，最高处位于西北部边缘，水深 4 205 m，山麓水深 5 527 m，最大高差约 1 322 m。 Located in the northeast of the Pacific Ocean, the seamount is 1,120 km northeast of the Musicians Seamounts. Being oval in shape as a whole, it is 12 km long and 10 km wide. In the middle of the seamount, it is high in the west and low in the east, with steep edges. The highest point, which is at the edge of the northwest of the flat top, is 4,205 m deep whereas the piedmont of the seamount is 5,527 m deep. Hence, the maximum total relief is about 1,322 m.		
专名释义 Reason for Choice of Name	该区域沿用音乐家相关命名体系群组化方法命名。魏良辅（1489—1566），字师召，号此斋，嘉靖年间杰出的戏曲音乐家、戏曲革新家。其对昆山腔的艺术发展有突出贡献，被后人奉为“昆曲之祖”，在曲艺界更有“曲圣”之称。故以魏良辅命名。 The features in this sea area are all named following grouping naming system related to musicians. There was a famous musician named Wei Liangfu (1489AD–1566AD), with the courtesy name Shizhao and literary name Cizhai. He was an outstanding musician and innovator of the Chinese traditional opera during the Jiajing period of the Ming Dynasty. Because of his outstanding contributions to the development of Kunshan (Kun) Opera, he was hailed as “Creator of Kunqu Opera” by later generations, and was known as “Sage of Opera” in the opera field. So the feature is named Weiliangfu Seamount after him.		

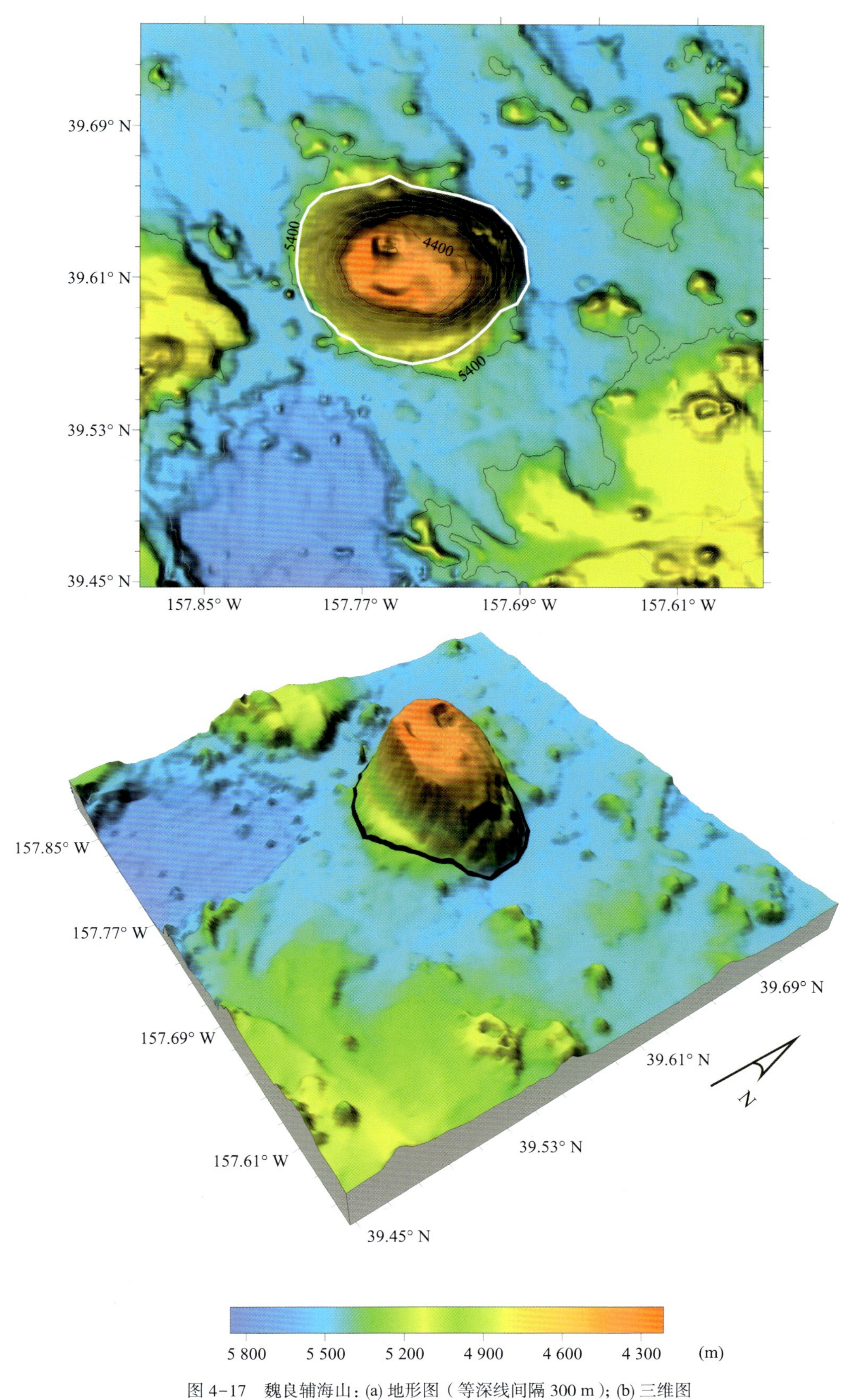

图 4-17 魏良辅海山：(a) 地形图（等深线间隔 300 m）；(b) 三维图

Fig.4-17 Weiliangfu Seamount: (a) Bathymetric map (Contours are in 300 m); (b) 3-D topographic map

4.2.17 编钟海丘群 Bianzhong Hills

中文名称 Chinese Name	Biānzhōng Hǎiqiūqún 编钟海丘群		
英文名称 English Name	Bianzhong Hills		
地理区域 / Location	东北太平洋 The Northeast Pacific Ocean		
特征点坐标 Coordinate	39°54.0′N 171°36.7′W 39°49.9′N 171°45.8′W	长度 / Length 宽度 / Width	56 km 20 km
水深 / Depth	4 864~6 021 m	高差 / Total Relief	1 157 m
发现情况 Discovery Facts	此海丘群于 2019 年“向阳红 19”船在执行航次调查时发现。 The hills were discovered in 2019 during the survey cruise carried out onboard the Chinese R/V *Xiang Yang Hong 19.*		
地形特征 Feature Description	编钟海丘群发育于太平洋东北部，位于音乐家海山群西北方向 1 200 km，由 2 个山体呈 SW-NE 方向排列组成，长和宽分别为 56 km 和 20 km。最高处位于东部海丘，峰顶水深 4 864m，两个较大海丘顶部相对平坦，周缘地势较陡，山麓水深 6 021 m，广泛分布多个规模较小的凸起。 Located in the northeast of the Pacific Ocean, the hills are 1,200 km northwest of the Musicians Seamounts. Consisting of two mountains, they extend in the SW-NE direction, and are 56 km long and 20 km wide. The highest point is located in the eastern hill, with a depth of 4,864 m at the peak. The top of the two larger hills is relatively flat whereas their surroundings are relatively steep. And they are 6,021 m deep at the piedmont, with many small protrusions which are widely distributed.		
专名释义 Reason for Choice of Name	该区域沿用中国乐器相关命名体系群组化方法命名。该海丘群呈线状分布，形似一组编钟，故以此命名。编钟是中国古代的一种大型打击乐器，兴起于西周，盛于春秋战国直至秦汉。编钟用青铜器铸成，按照音律高低次序依次排列悬挂于巨大的钟架上，用丁字木槌或长棒敲打铜钟，可以演奏出美妙的乐曲。 The features in this sea area are all named following the ancient Chinese musical instrument naming system. There was a famous ancient musical instrument called Bianzhong. The Bianzhong, a large set of percussion instruments in ancient China, came into being during the Western Zhou Dynasty, and became popular from the Spring and Autumn Period and the Warring States Period to the Qin and Han Dynasties. Made of bronze, Bianzhong are suspended on a huge bell-cot in sequence according to the order of pitches and tones. By striking the bronze bells with T-shaped wooden hammers or long sticks, people can play beautiful music through Bianzhong. The hills are distributed in a linear manner, resembling a set of Bianzhong. Hence, the features are named Bianzhong Hills.		

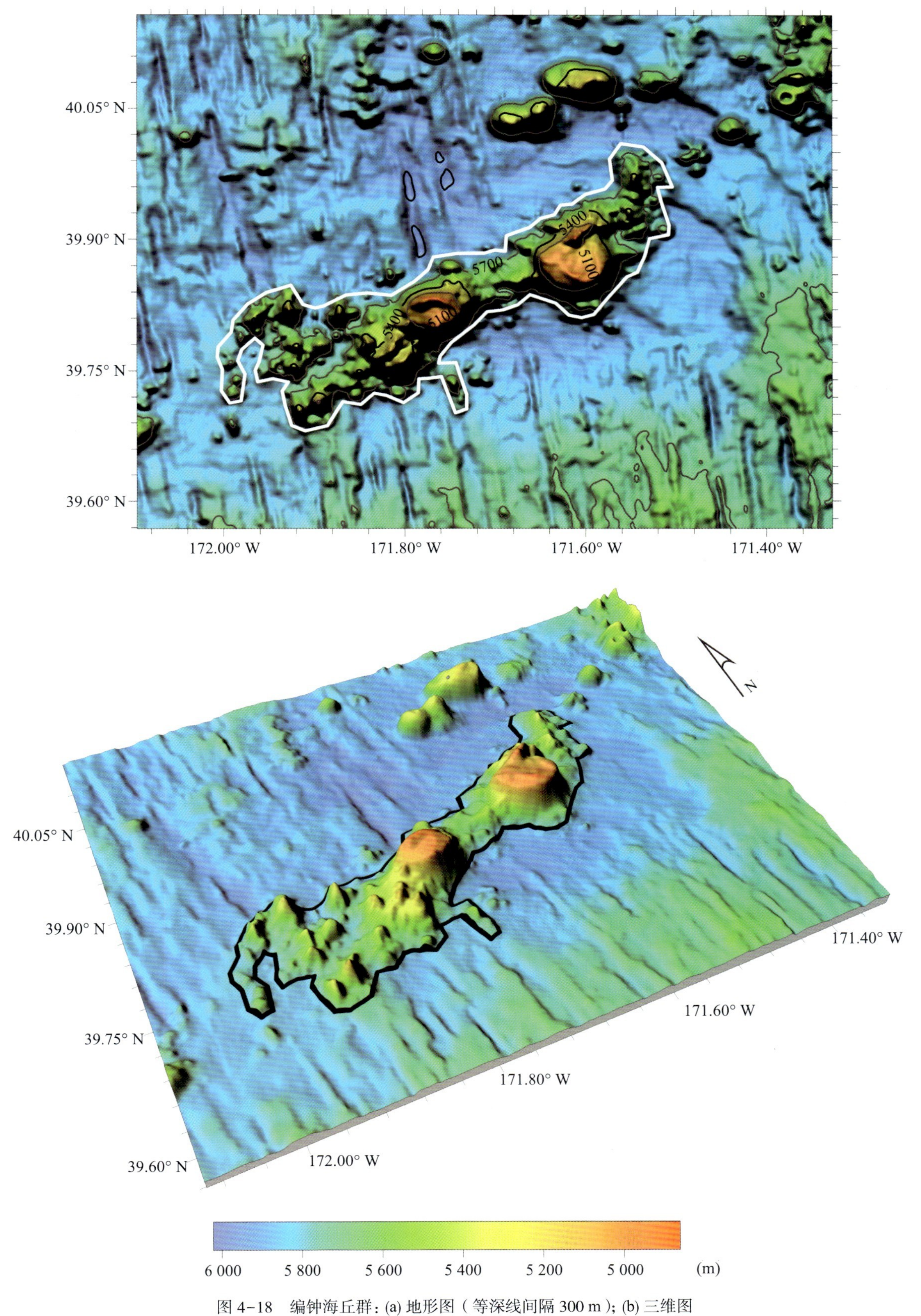

图 4-18 编钟海丘群：(a) 地形图（等深线间隔 300 m）；(b) 三维图

Fig.4-18 Bianzhong Hills: (a) Bathymetric map (Contours are in 300 m); (b) 3-D topographic map

4.2.18 葫芦埙海丘群 Huluxun Hills

<table>
<tr><td>中文名称
Chinese Name</td><td colspan="3">Húluxūn Hǎiqiūqún
葫芦埙海丘群</td></tr>
<tr><td>英文名称
English Name</td><td colspan="3">Huluxun Hills</td></tr>
<tr><td>地理区域 / Location</td><td colspan="3">东北太平洋 The Northeast Pacific Ocean</td></tr>
<tr><td rowspan="2">特征点坐标
Coordinate</td><td rowspan="2">40°05.2′N 170°38.9′W
40°03.8′N 170°36.9′W
40°06.6′N 170°29.0′W</td><td>长度 / Length</td><td>33 km</td></tr>
<tr><td>宽度 / Width</td><td>10 km</td></tr>
<tr><td>水深 / Depth</td><td>4 873~5 889 m</td><td>高差 / Total Relief</td><td>1 016 m</td></tr>
<tr><td>发现情况
Discovery Facts</td><td colspan="3">此海丘群于 2019 年“向阳红 19”船在执行航次调查时发现。
The hills were discovered in 2019 during the survey cruise carried out onboard the Chinese R/V Xiang Yang Hong 19.</td></tr>
<tr><td>地形特征
Feature Description</td><td colspan="3">葫芦埙海丘群发育于太平洋东北部，位于音乐家海山群东北方向 1 120 km，由 3 个山体呈 NE-SW 向排列组成，长和宽分别为 33 km 和 10 km。最高处位于东部海山，峰顶水深 4 873 m，底部水深 5 889 m，最大高差约 1 016 m。
Located in the northeast of the Pacific Ocean, the hills are 1,120 km northeast of the Musicians Seamounts. Consisting of three mountains extending in the NE-SW direction, they are 33 km long and 10 km wide. The highest point is located in the eastern seamount, with a depth of 4,873 m at the peak, and a depth of 5,889 m at the foot. The maximum total relief is about 1,016 m.</td></tr>
<tr><td>专名释义
Reason for Choice of Name</td><td colspan="3">该区域沿用中国乐器相关命名体系群组化方法命名。葫芦埙多为陶土烧制的一种吹奏乐器，汉族特有，外形像葫芦，吹奏音色柔和优美，该海丘群形似我国吹奏乐器葫芦埙，故以此义命名。
The features in this area are all named following the ancient Chinese musical instrument naming system. There is an ancient Chinese musical instrument called Huluxun which is unique to the Han nationality, is generally a kind of wind instrument made of clay. It looks like a gourd and produces soft and elegant sounds. The hills are quite like the wind instrument Huluxun. Hence, the features are named Huluxun Hills.</td></tr>
</table>

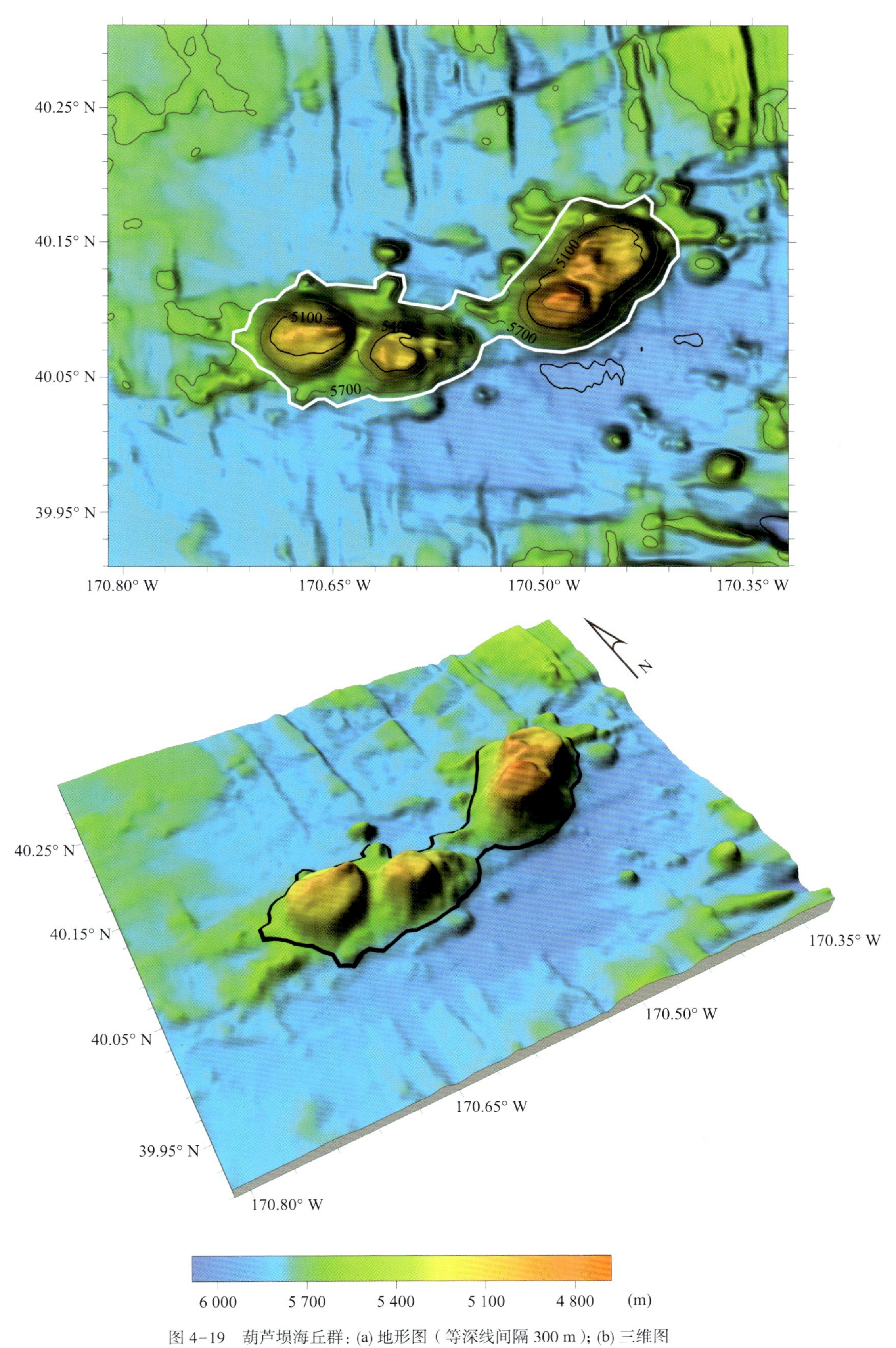

图 4-19 葫芦埙海丘群：(a) 地形图（等深线间隔 300 m）；(b) 三维图

Fig.4-19 Huluxun Hills: (a) Bathymetric map (Contours are in 300 m); (b) 3-D topographic map

4.2.19 箜篌海脊群 Konghou Ridges

<table>
<tr><td>中文名称
Chinese Name</td><td colspan="3">Kōnghóu Hǎijǐqún
箜篌海脊群</td></tr>
<tr><td>英文名称
English Name</td><td colspan="3">Konghou Ridges</td></tr>
<tr><td>地理区域 / Location</td><td colspan="3">东北太平洋 The Northeast Pacific Ocean</td></tr>
<tr><td rowspan="2">特征点坐标
Coordinate</td><td rowspan="2">39°12.2′N 170°41.3′W
39°20.5′N 170°59.6′W
39°18.7′N 171°01.3′W</td><td>长度 / Length</td><td>72 km</td></tr>
<tr><td>宽度 / Width</td><td>49 km</td></tr>
<tr><td>水深 / Depth</td><td>3 949~6 003 m</td><td>高差 / Total Relief</td><td>2 054 m</td></tr>
<tr><td>发现情况
Discovery Facts</td><td colspan="3">此海脊群于 2019 年“向阳红 19”船在执行航次调查时发现。
The ridges were discovered in 2019 during the survey cruise carried out onboard the Chinese R/V Xiang Yang Hong 19.</td></tr>
<tr><td>地形特征
Feature Description</td><td colspan="3">箜篌海脊群发育于太平洋东北部，位于音乐家海山群西北 1 000 km，近 NW–SE 向展布，由 3 条海脊组成，长和宽分别为 72 km 和 49 km。最高处位于西部海山，顶部水深 3 949 m，底部水深 6 003 m，最大高差约 2 054 m。
Located in the northeast of the Pacific Ocean, the ridges are 1,000 km northwest of the Musicians Seamounts. Consisting of 3 ridges, they extend in the NW-SE direction, and are 72 km long and 49 km wide. The highest point is located in the western seamount, which is 3,949 m deep at the top and 6,003 m deep at the foot, with a maximum total relief of about 2,054 m.</td></tr>
<tr><td>专名释义
Reason for Choice of Name</td><td colspan="3">该区域沿用中国乐器相关命名体系群组化方法命名。箜篌，是中国古代传统的弹弦乐器，最早可追溯到汉代，其音色如溪流般轻柔，音域如江河般宽广。该海脊群形似敦煌壁画中一只倒置的凤首箜篌，故以此命名。
The features in this sea area are all named following the ancient Chinese musical instrument naming system. There is an ancient Chinese musical instrument called Konghou which can be traced back to the Han Dynasty. It is a traditional plucked string instrument of ancient China. Its tone is as gentle as a stream and its range is as wide as a river. The ridges look like an upside-down phoenix-headed Konghong in Dunhuang Frescoes. Hence, the features are named Konghou Ridges.</td></tr>
</table>

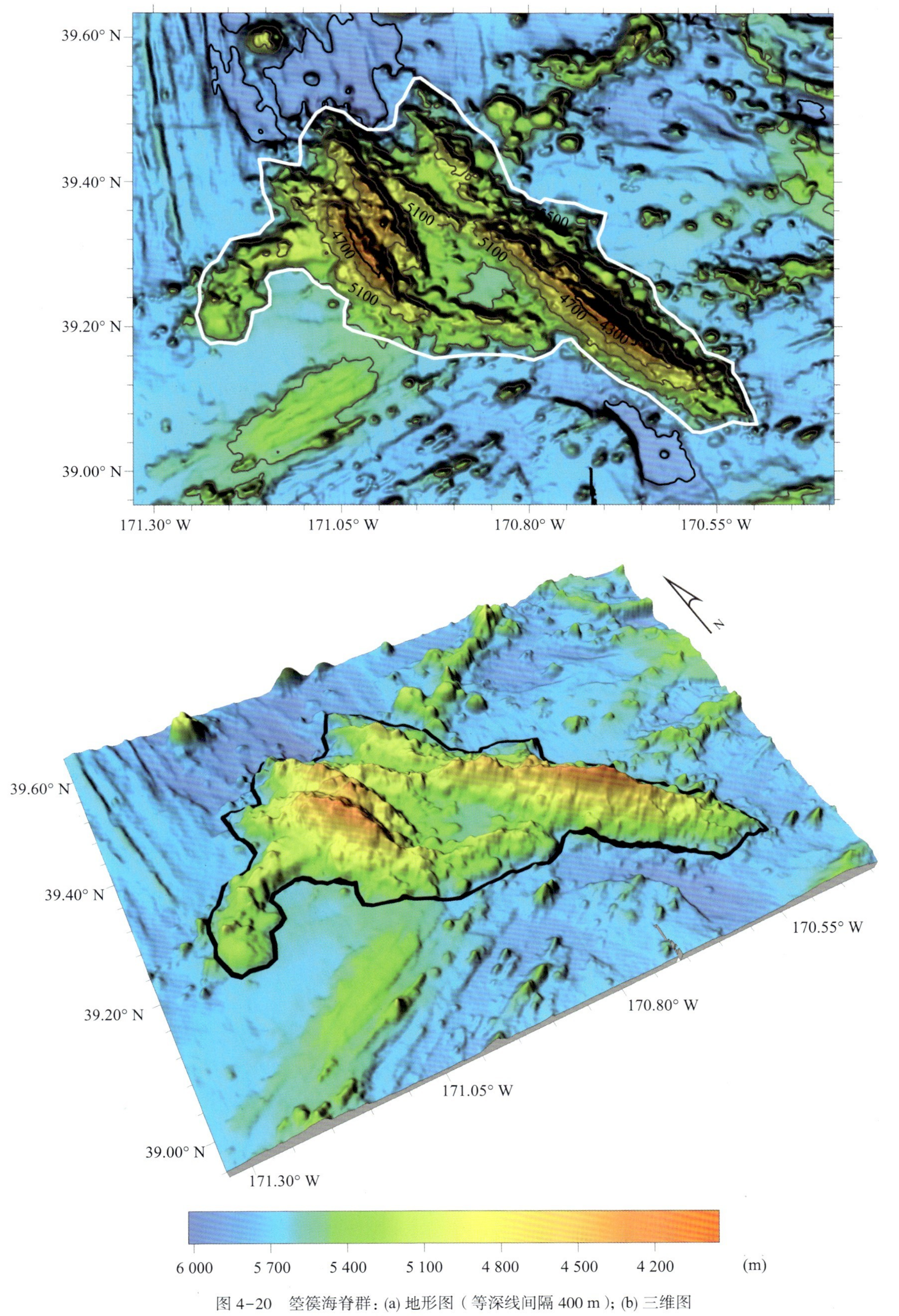

图 4-20 箜篌海脊群：(a) 地形图（等深线间隔 400 m）；(b) 三维图

Fig.4-20 Konghou Ridges: (a) Bathymetric map (Contours are in 400 m); (b) 3-D topographic map

4.2.20 芦笙海渊 Lusheng Deep

<table>
<tr><td>中文名称
Chinese Name</td><td colspan="3">Lúshēng Hǎiyuān
芦笙海渊</td></tr>
<tr><td>英文名称
English Name</td><td colspan="3">Lusheng Deep</td></tr>
<tr><td>地理区域 / Location</td><td colspan="3">东北太平洋 The Northeast Pacific Ocean</td></tr>
<tr><td rowspan="2">特征点坐标
Coordinate</td><td rowspan="2">39°26.2′N 169°45.0′W</td><td>长度 / Length</td><td>76 km</td></tr>
<tr><td>宽度 / Width</td><td>25 km</td></tr>
<tr><td>水深 / Depth</td><td>5 643~6 971 m</td><td>高差 / Total Relief</td><td>1 328 m</td></tr>
<tr><td>发现情况
Discovery Facts</td><td colspan="3">此海渊于 2019 年“向阳红 19”船在执行航次调查时发现。
The deep was discovered in 2019 during the survey cruise carried out onboard the Chinese R/V Xiang Yang Hong 19 .</td></tr>
<tr><td>地形特征
Feature Description</td><td colspan="3">芦笙海渊发育于太平洋东北部，位于音乐家海山群西北 1 000 km，呈 E-W 向爪型展布，长和宽分别为 76 km 和 25 km。最浅处水深 5 643 m，最深处水深 6 971 m，最大高差约 1 328 m。该海渊内部沟脊相间，在西北部发散，汇聚于东部，周缘较为陡峭。
Located in the northeast of the Pacific Ocean, the deep is 1,000 km northwest of the Musicians Seamounts. Extending in the E-W direction in the shape of a claw, it is 76 km long and 25 km wide. The shallowest point is 5,643 m deep, and the deepest point is 6,971 m deep, with a maximum total relief of about 1,328 m. Trenches and ridges which alternate with each other within the deep are radiated from the northwest and converged in the east, with steep edges around.</td></tr>
<tr><td>专名释义
Reason for Choice of Name</td><td colspan="3">该区域沿用中国乐器相关命名体系群组化方法命名。芦笙是中国西南地区苗族、瑶族、侗族等民族的簧管乐器，其前身为竽，最早可追溯至宋代。其音调多样，气势恢宏，独奏时清雅柔和，合奏时高昂激越。该海渊沟脊相间，形似芦笙，故以此命名。
The features in this sea area are all named following the Chinese musical instrument naming system. The Lusheng is a reed instrument of ethnic groups such as Miao, Yao, and Dong in Southwest China. Its predecessor is Yu, which can be traced back to the Song Dynasty. With diverse and magnificent tones, it produces elegant and soft tunes when played independently, and passionate and exciting tunes when played with other instruments. With trenches and ridges alternating with each other, the deep looks like a Lusheng. Hence, the feature is named Lusheng Deep.</td></tr>
</table>

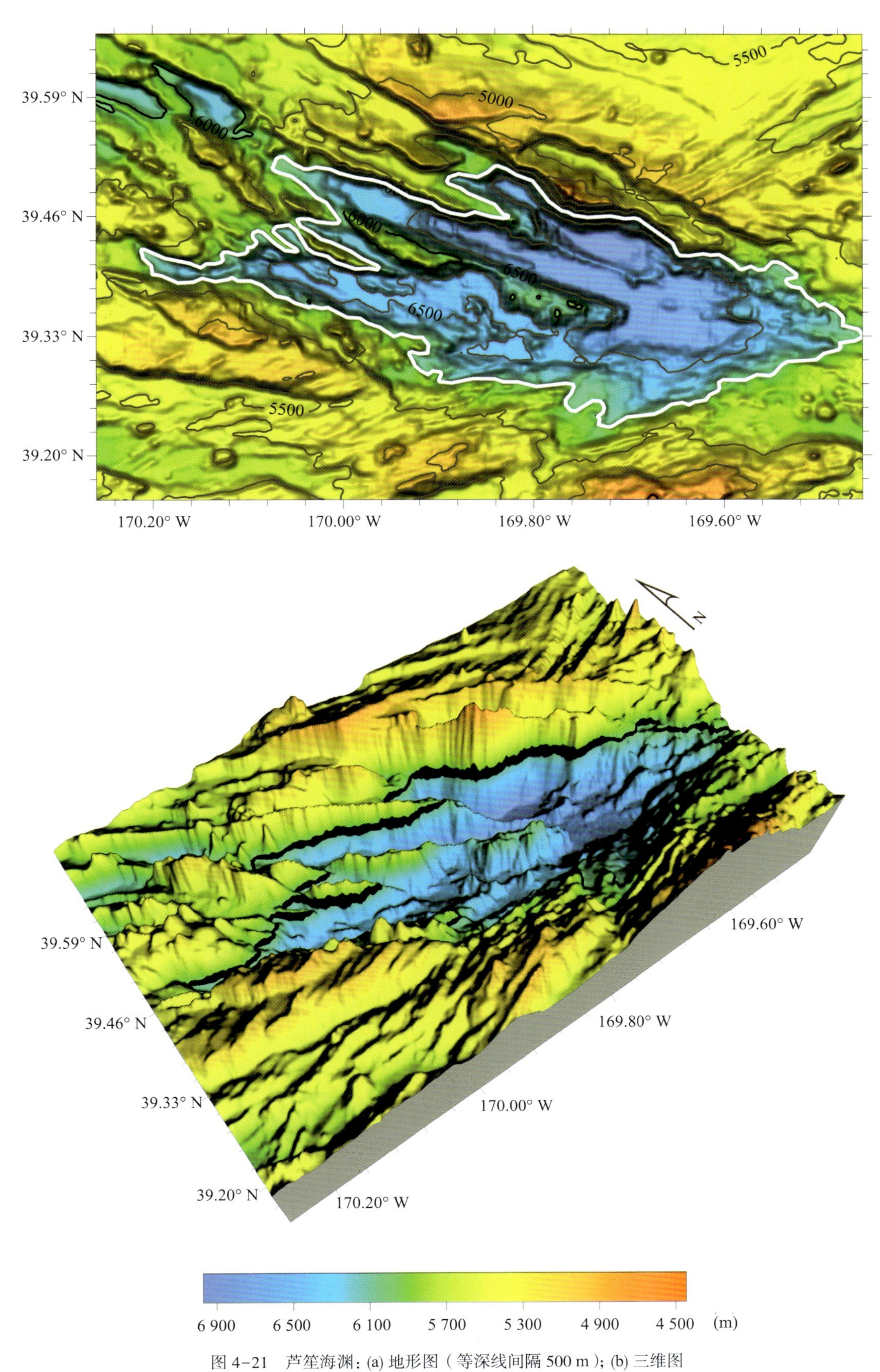

图 4-21 芦笙海渊：(a) 地形图（等深线间隔 500 m）；(b) 三维图

Fig.4-21 Lusheng Deep: (a) Bathymetric map (Contours are in 500 m); (b) 3-D topographic map

4.2.21 梆笛海脊 Bangdi Ridge

中文名称 Chinese Name	Bāngdí Hǎijǐ 梆笛海脊		
英文名称 English Name	Bangdi Ridge		
地理区域 / Location	东北太平洋 The Northeast Pacific Ocean		
特征点坐标 Coordinate	40°23.3′N 169°21.3′W	长度 / Length	59 km
		宽度 / Width	17 km
水深 / Depth	4 361~5 806 m	高差 / Total Relief	1 445 m
发现情况 Discovery Facts	此海脊于 2019 年“向阳红 19”船在执行航次调查时发现。 The ridge was discovered in 2019 during the survey cruise carried out onboard the Chinese R/V *Xiang Yang Hong 19.*		
地形特征 Feature Description	梆笛海脊群发育于太平洋东北部，位于音乐家海山群西北 1 100 km，呈 NE–SW 向带状分布，长和宽分别为 59 km 和 17 km。顶部水深 4 361 m，底部水深 5 806 m，最大高差约 1 445 m。 Located in the northeast of the Pacific Ocean, the ridge is 1,100 km northwest of the Musicians Seamounts. Extending in the NE-SW direction, it is 59 km long and 17 km wide. Being 4,361 m deep at the top and 5,806 m deep at the foot, the ridge has a maximum total relief of about 1,445 m.		
专名释义 Reason for Choice of Name	该区域沿用中国乐器相关命名体系群组化方法命名。梆笛，是中国传统音乐中常用的横吹木管乐器之一，其音调高亢、音色明亮，是我国北方吹歌会、评剧和梆子戏曲的主要伴奏。该海脊形体修长短小，形似梆笛，与曲笛海脊一北一南分布，故以梆笛命名。 The features in this sea area are all named following the Chinese musical instrument naming system. The Bangdi is one of the horizontally blown woodwind instruments commonly used in the traditional Chinese music. It has a high-pitched and bright tone, and serves as a major accompanying instrument for the ensemble of Chinese Wind and Percussion Opera, Ping Opera and Bangzi Opera in North China. This ridge is slender and short, resembling a Bangdi. It is located in the north whereas Qudi Ridge is located in the south. Hence, the feature is named Bangdi Ridge.		

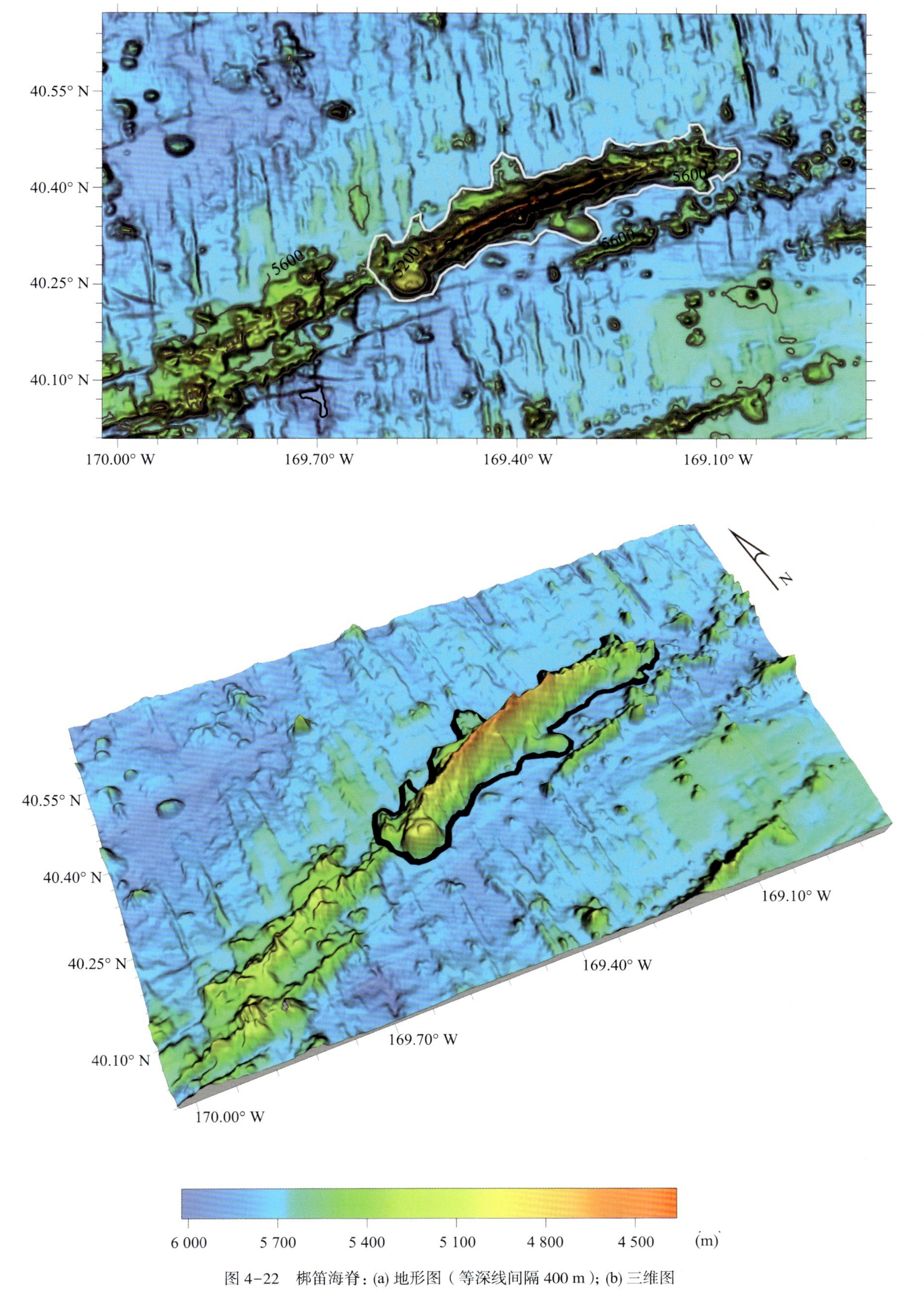

图 4-22 梆笛海脊：(a) 地形图（等深线间隔 400 m）；(b) 三维图

Fig.4-22 Bangdi Ridge: (a) Bathymetric map (Contours are in 400 m); (b) 3-D topographic map

4.2.22 鸿鹄海脊 Honghu Ridge

<table>
<tr><td>中文名称
Chinese Name</td><td colspan="3">Hónghú Hǎijǐ
鸿鹄海脊</td></tr>
<tr><td>英文名称
English Name</td><td colspan="3">Honghu Ridge</td></tr>
<tr><td>地理区域 / Location</td><td colspan="3">东北太平洋 The Northeast Pacific Ocean</td></tr>
<tr><td rowspan="2">特征点坐标
Coordinate</td><td rowspan="2">39°04.9′N 168°27.1′W</td><td>长度 / Length</td><td>214 km</td></tr>
<tr><td>宽度 / Width</td><td>48 km</td></tr>
<tr><td>水深 / Depth</td><td>2 097~6 031 m</td><td>高差 / Total Relief</td><td>3 934 m</td></tr>
<tr><td>发现情况
Discovery Facts</td><td colspan="3">此海脊于 2019 年“向阳红 19”船在执行航次调查时发现。
The ridge was discovered in 2019 during the survey cruise carried out onboard the Chinese R/V Xiang Yang Hong 19.</td></tr>
<tr><td>地形特征
Feature Description</td><td colspan="3">鸿鹄海脊发育于太平洋东北部，位于音乐家海山群西北 900km，近 NW-SE 向展布，长和宽分别为 214 km 和 48 km，发育一小规模平顶海山。该海脊顶部水深 2 097 m，山麓水深 6 031 m，最大高差约 3 934 m。
Located in the northeast of the Pacific Ocean, the ridge is 900 km northwest of the Musicians Seamounts, and extends in the NW-SE direction. It is 214 km long and 48 km wide. The ridge is 2,097 m deep at the top, and 6,031 m deep at the foot, with a maximum total relief of about 3,934 m.</td></tr>
<tr><td>专名释义
Reason for Choice of Name</td><td colspan="3">鸿鹄是我国古人对大雁、天鹅之类飞行极为高远鸟类的通称，常用来比喻志向远大的人。该海脊形似一只展开双翅的大雁，翱翔在海底，故以鸿鹄命名。
The Honghu is a general ancient term for wild geese, swans and other birds that fly extremely high. It is often used to refer to the highly ambitious people. The ridge looks like a wild goose soaring over the seabed with wings spread out. Hence ,the feature is named Honghu Ridge.</td></tr>
</table>

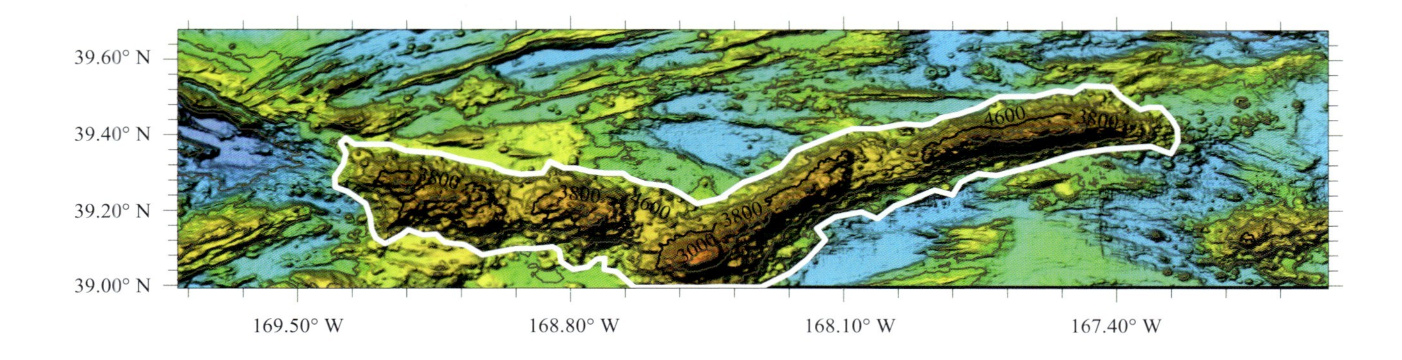

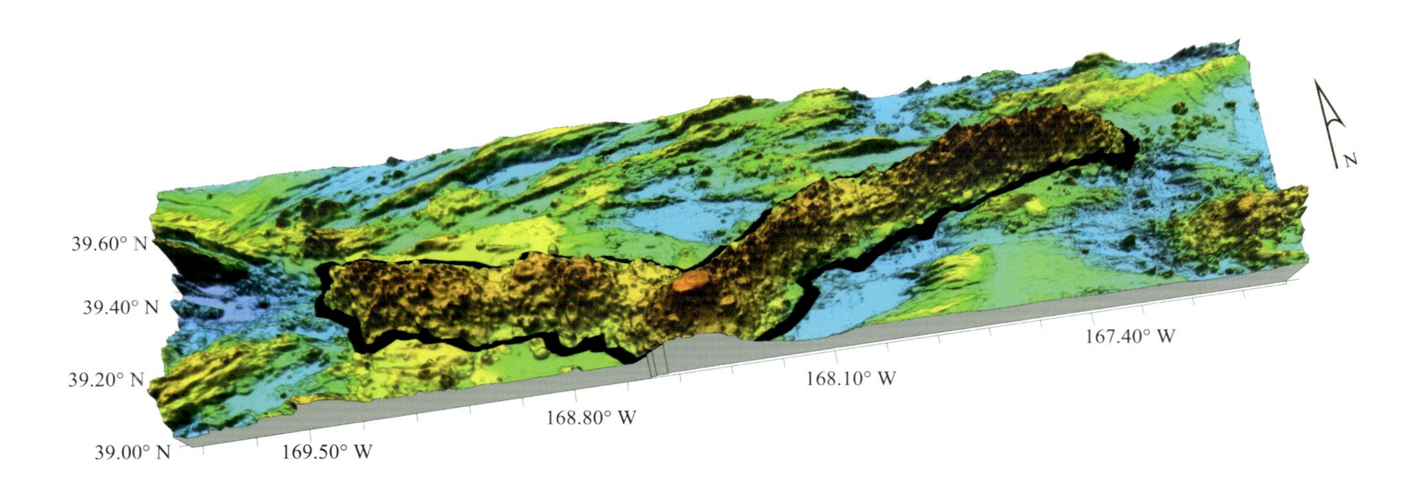

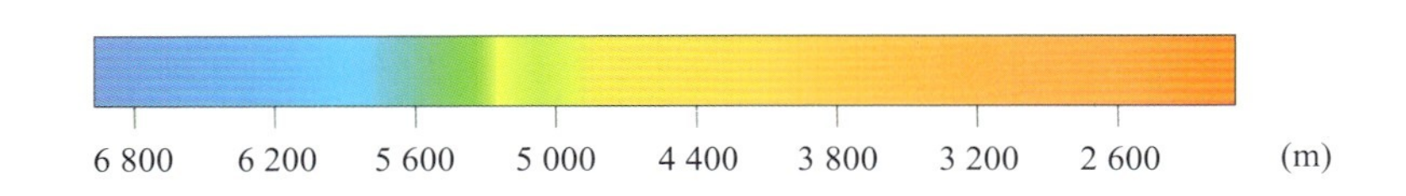

图 4-23 鸿鹄海脊：(a) 地形图（等深线间隔 800 m）；(b) 三维图

Fig.4-23 Honghu Ridge: (a) Bathymetric map (Contours are in 800 m); (b) 3-D topographic map

4.2.23 铜钹圆丘 Tongbo Knoll

<table>
<tr><td>中文名称
Chinese Name</td><td colspan="4">Tóngbó Yuánqiū
铜钹圆丘</td></tr>
<tr><td>英文名称
English Name</td><td colspan="4">Tongbo Knoll</td></tr>
<tr><td>地理区域 / Location</td><td colspan="4">东北太平洋 The Northeast Pacific Ocean</td></tr>
<tr><td rowspan="2">特征点坐标
Coordinate</td><td rowspan="2">40°56.9′N 167°38.2′W</td><td>长度 / Length</td><td colspan="2">15 km</td></tr>
<tr><td>宽度 / Width</td><td colspan="2">10 km</td></tr>
<tr><td>水深 / Depth</td><td>5 005~5 958 m</td><td>高差 / Total Relief</td><td colspan="2">953 m</td></tr>
<tr><td>发现情况
Discovery Facts</td><td colspan="4">此圆丘于 2019 年“向阳红 06”船在执行航次调查时发现。
The knoll was discovered in 2019 during the survey cruise carried out onboard the Chinese R/V Xiang Yang Hong 06.</td></tr>
<tr><td>地形特征
Feature Description</td><td colspan="4">铜钹圆丘发育于太平洋东北部，位于音乐家海山群西北 1120km，整体呈椭圆状，长和宽分别为 15 km 和 10 km。该海丘顶部水深 5 005 m，底部水深 5 958 m，最大高差约 953 m。
Located in the northeast of the Pacific Ocean, the knoll is 1,120 km northwest of the Musicians Seamounts. Being oval in shape as a whole, it is 15 km long and 10 km wide. The hill is 5,005 m deep at the top and 5,958 m deep at the foot, with a maximum total relief of about 953 m.</td></tr>
<tr><td>专名释义
Reason for Choice of Name</td><td colspan="4">该区域沿用中国乐器相关命名体系群组化方法命名。铜钹，打击乐器，用响铜所造，其形如圆盘，中央隆起如帽，声音洪亮、浑厚。该圆丘主体浑圆，形似铜钹，故以铜钹命名。
The features in this sea area are all named following the Chinese musical instrument naming system. The Tongbo is a percussion instrument made of cacophony (an alloy which is a mixture of copper, lead and tin). The instrument is shaped like a disc, with an elevation in the center like a hat. It produces loud and thick sounds. The main body of the knoll is rounded, and it looks like a Tongbo. Hence, the feature is named Tongbo Knoll.</td></tr>
</table>

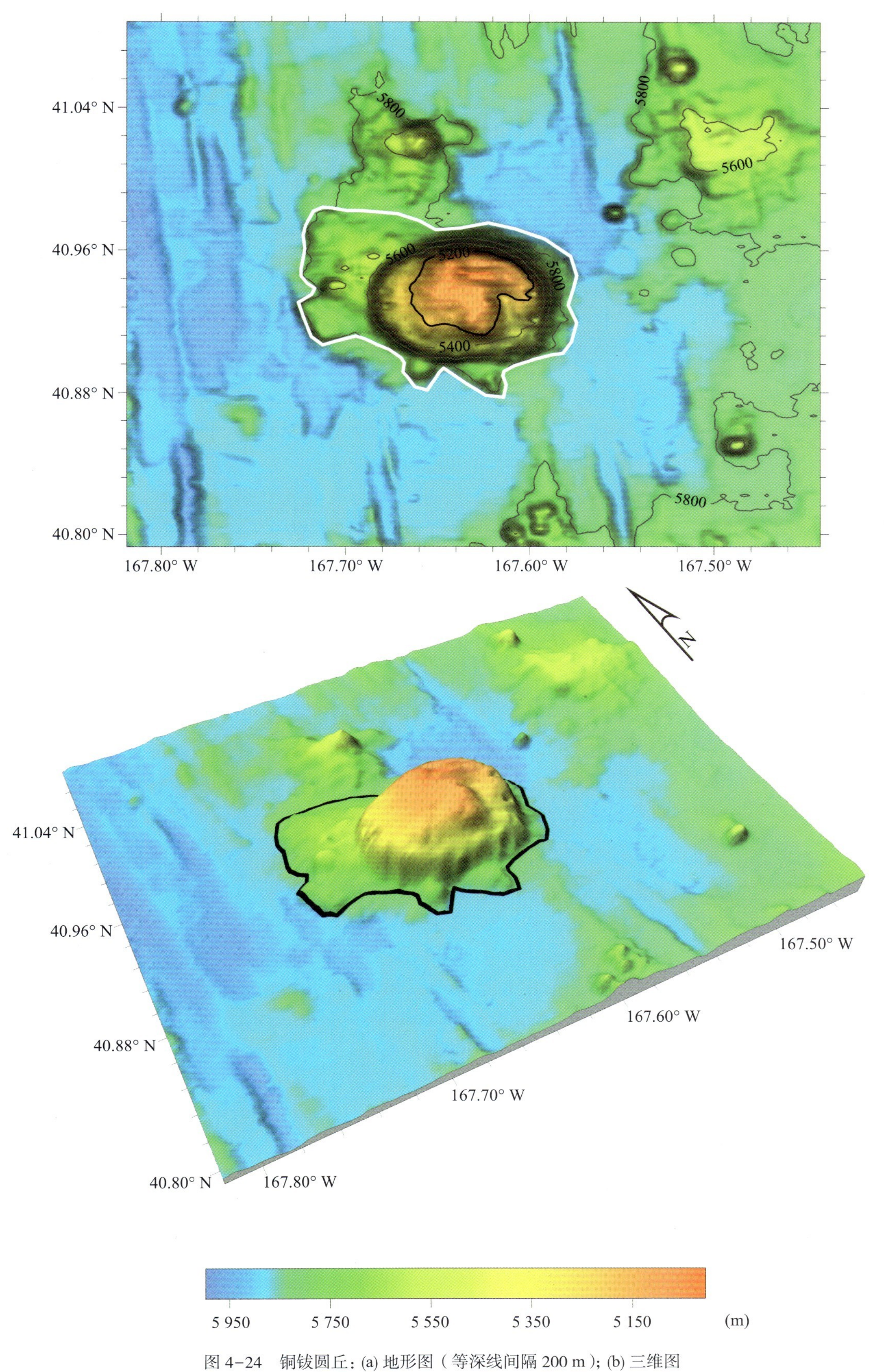

图 4-24 铜钹圆丘：(a) 地形图（等深线间隔 200 m）；(b) 三维图
Fig.4-24 Tongbo Knoll: (a) Bathymetric map (Contours are in 200 m); (b) 3-D topographic map

4.2.24 曲笛海脊 Qudi Ridge

<table>
<tr><td>中文名称
Chinese Name</td><td colspan="3">Qǔdí Hǎijǐ
曲笛海脊</td></tr>
<tr><td>英文名称
English Name</td><td colspan="3">Qudi Ridge</td></tr>
<tr><td>地理区域 / Location</td><td colspan="3">东北太平洋 The Northeast Pacific Ocean</td></tr>
<tr><td rowspan="2">特征点坐标
Coordinate</td><td rowspan="2">39°16.9′N 166°32.7′W</td><td>长度 / Length</td><td>140 km</td></tr>
<tr><td>宽度 / Width</td><td>23 km</td></tr>
<tr><td>水深 / Depth</td><td>3 516~5 981 m</td><td>高差 / Total Relief</td><td>2 465 m</td></tr>
<tr><td>发现情况
Discovery Facts</td><td colspan="3">此海脊于 2019 年“向阳红 06”船在执行航次调查时发现。
The ridge was discovered in 2019 during the survey cruise carried out onboard the Chinese R/V Xiang Yang Hong 06.</td></tr>
<tr><td>地形特征
Feature Description</td><td colspan="3">曲笛海脊发育于太平洋东北部，位于音乐家海山群西北 855 km，呈 NE-SW 向长条状展布，长和宽分别为 140 km 和 23 km。该海脊背部高耸，水深 3 516 m，侧翼陡峭，底部水深 5 981 m，最大高差约 2 465 m。
Located in the northeast of the Pacific Ocean, the ridge is 855 km northwest of the Musicians Seamounts. Extending in the NE-SW direction like a long stripe, the ridge is 140 km long and 23 km wide. The back of the ridge is towering with a depth of 3,516 m, whereas the flanks are steep, with a depth of 5,981 m at the foot. The maximum total relief is about 2,465 m.</td></tr>
<tr><td>专名释义
Reason for Choice of Name</td><td colspan="3">该区域沿用中国乐器相关命名体系群组化方法命名。曲笛，是中国传统音乐中常用的横吹木管乐器之一，其音色淳厚柔和、清新圆润，因伴奏我国南方昆曲而得名。该海脊形体修长，与梆笛海脊一南一北分布，故以曲笛命名。
The features in this sea area are all named following the Chinese musical instrument naming system. The Qudi is one of the horizontally blown woodwind instruments commonly used in the traditional Chinese music, with a pure, soft, fresh and round tone. It is named after its accompaniment to the Kunqu Opera in South China. The feature is a slender ridge located in the south whereas Bangdi Ridge is located in the north. Hence, the feature is named Qudi Ridge.</td></tr>
</table>

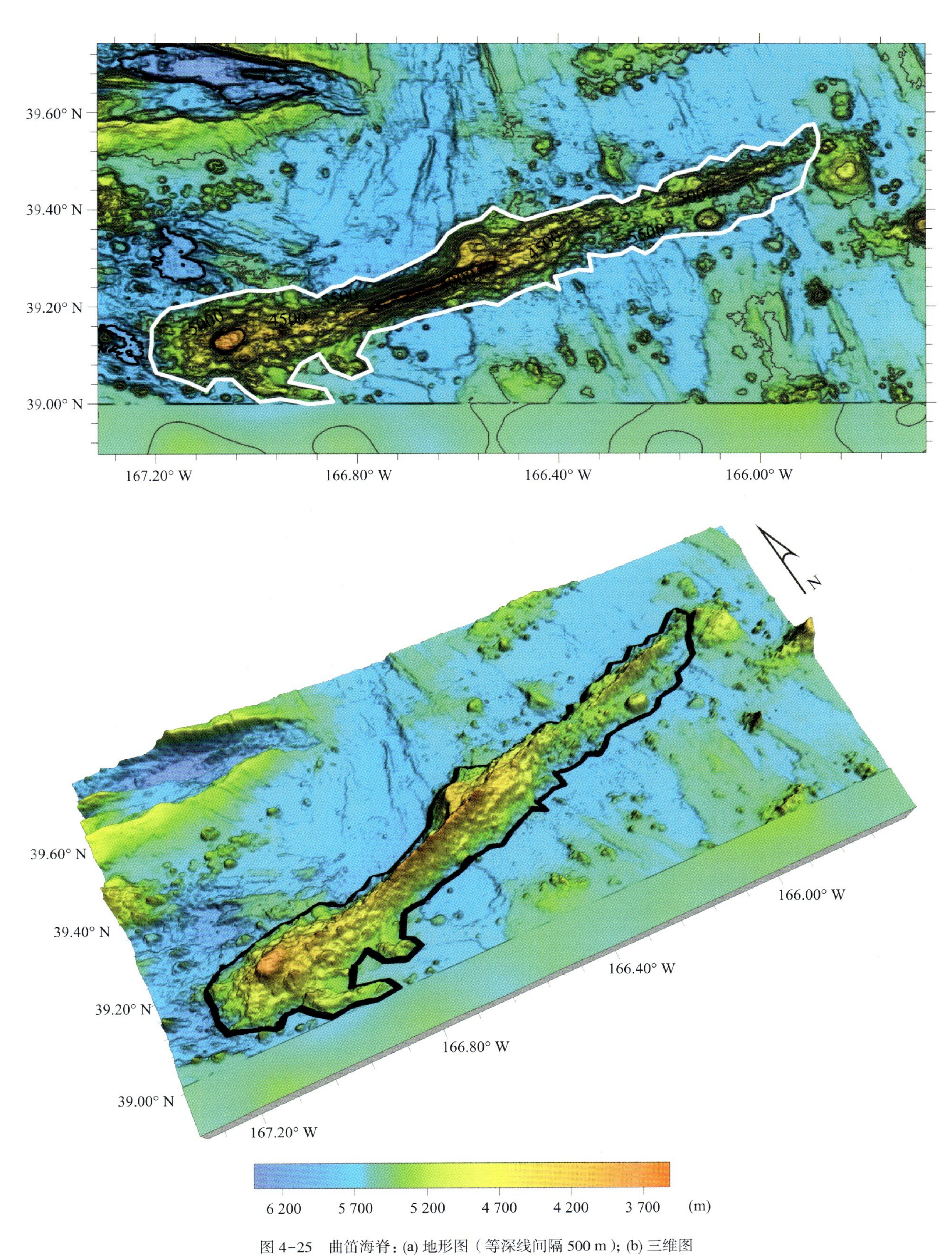

图 4-25 曲笛海脊：(a) 地形图（等深线间隔 500 m）；(b) 三维图

Fig.4-25 Qudi Ridge: (a) Bathymetric map (Contours are in 500 m); (b) 3-D topographic map

4.2.25 铜锣圆丘 Tongluo Knoll

中文名称 Chinese Name	Tóngluó Yuánqiū 铜锣圆丘		
英文名称 English Name	Tongluo Knoll		
地理区域 / Location	东北太平洋 The Northeast Pacific Ocean		
特征点坐标 Coordinate	39°21.0′N 165°23.6′W	长度 / Length	15 km
		宽度 / Width	10 km
水深 / Depth	4 803~5 658 m	高差 / Total Relief	855 m
发现情况 Discovery Facts	此圆丘于 2019 年"向阳红 06"船在执行航次调查时发现。 The knoll was discovered in 2019 during the survey cruise carried out onboard the Chinese R/V *Xiang Yang Hong 06.*		
地形特征 Feature Description	铜锣圆丘发育于太平洋东北部，位于音乐家海山群以北 845 km，长和宽分别为 15 km 和 10 km。该圆丘顶部水深 4 803 m，底部水深 5 658 m，最大高差约 855 m。圆丘顶部东西低中间高，周缘较为陡峭。 Located in the northeast of the Pacific Ocean, the knoll is 845 km north of the Musicians Seamounts. It is 15 km long and 10 km wide. The knoll is 4,803 m deep at the top and 5,658 m deep at the foot, with a maximum total relief of about 855 m. At the top of knoll, it is tall in the middle, low in the east and west, and steep on the periphery.		
专名释义 Reason for Choice of Name	该区域沿用中国乐器相关命名体系群组化方法命名。铜锣，由青铜制成，锣面呈圆盘形，中央微鼓起，是中国传统响器，其音色柔和、洪亮，声音穿透力强。该圆丘主体浑圆，形似铜锣，故以铜锣命名。 The features in this area are all named flowing the Chinese musical instrument naming system. The Tongluo, made of bronze, has a disc-shaped surface and a small bulge in the center. It is a traditional Chinese musical instrument with a soft, resonant tone and strong sound penetration. The main body of the knoll is rounded, which looks like a Tongluo. Hence, the knoll is named Tongluo Knoll.		

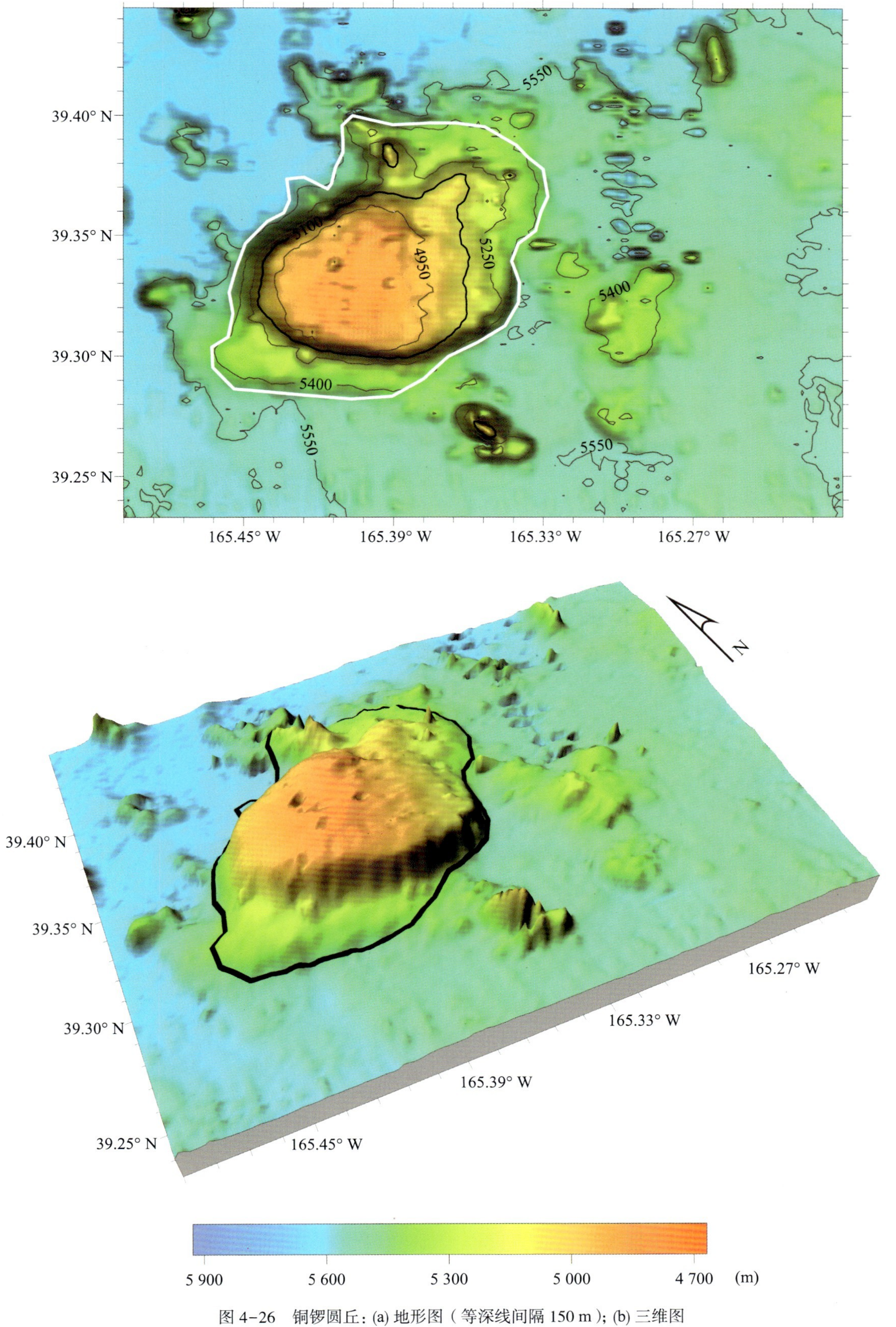

图 4-26 铜锣圆丘：(a) 地形图（等深线间隔 150 m）；(b) 三维图

Fig.4-26 Tongluo Knoll: (a) Bathymetric map (Contours are in 150 m); (b) 3-D topographic map

4.2.26 船锚海山 Chuanmao Seamount

中文名称 Chinese Name	Chuánmáo Hǎishān 船锚海山		
英文名称 English Name	Chuanmao Seamount		
地理区域 / Location	东北太平洋 The Northeast Pacific Ocean		
特征点坐标 Coordinate	41°00.9′N 163°48.8′W	**长度 / Length**	53 km
		宽度 / Width	31 km
水深 / Depth	4 317~5 871 m	**高差 / Total Relief**	1 554 m
发现情况 Discovery Facts	此海山于 2020 年“向阳红 01”船在执行航次调查时发现。 The seamount was discovered in 2020 during the survey cruise carried out onboard the Chinese R/V *Xiang Yang Hong 01*.		
地形特征 Feature Description	船锚海山发育于太平洋东北部，位于门多西诺断裂带以北 540 km，呈 E-W 走向，长和宽分别为 53 km 和 31 km，正北发育长约 30 km 的山脊。该海山最高处位于西部山顶，水深 4 317 m，山麓水深 5 871 m，最大高差约 1 554 m。 Located in the northeast of the Pacific Ocean, the seamount is 540 km north of the Mendocino Fracture Zone. Extending in the E-W direction, it is 53 km long and 31 km wide. In the north, there is a ridge, which is about 30 km long. The highest point of the seamount is located at the summit in the west, which is 4,317 m deep at the top, and 5,871 m deep at the piedmont, with a maximum total relief of about 1,554 m.		
专名释义 Reason for Choice of Name	该海山的三个山脊形似船锚的爪型，故以船锚命名。 The three ridges of this feature are shaped like a claw-shaped boat anchor (Chuanmao in Chinese). Hence, the feature is named Chuanmao Seamount.		

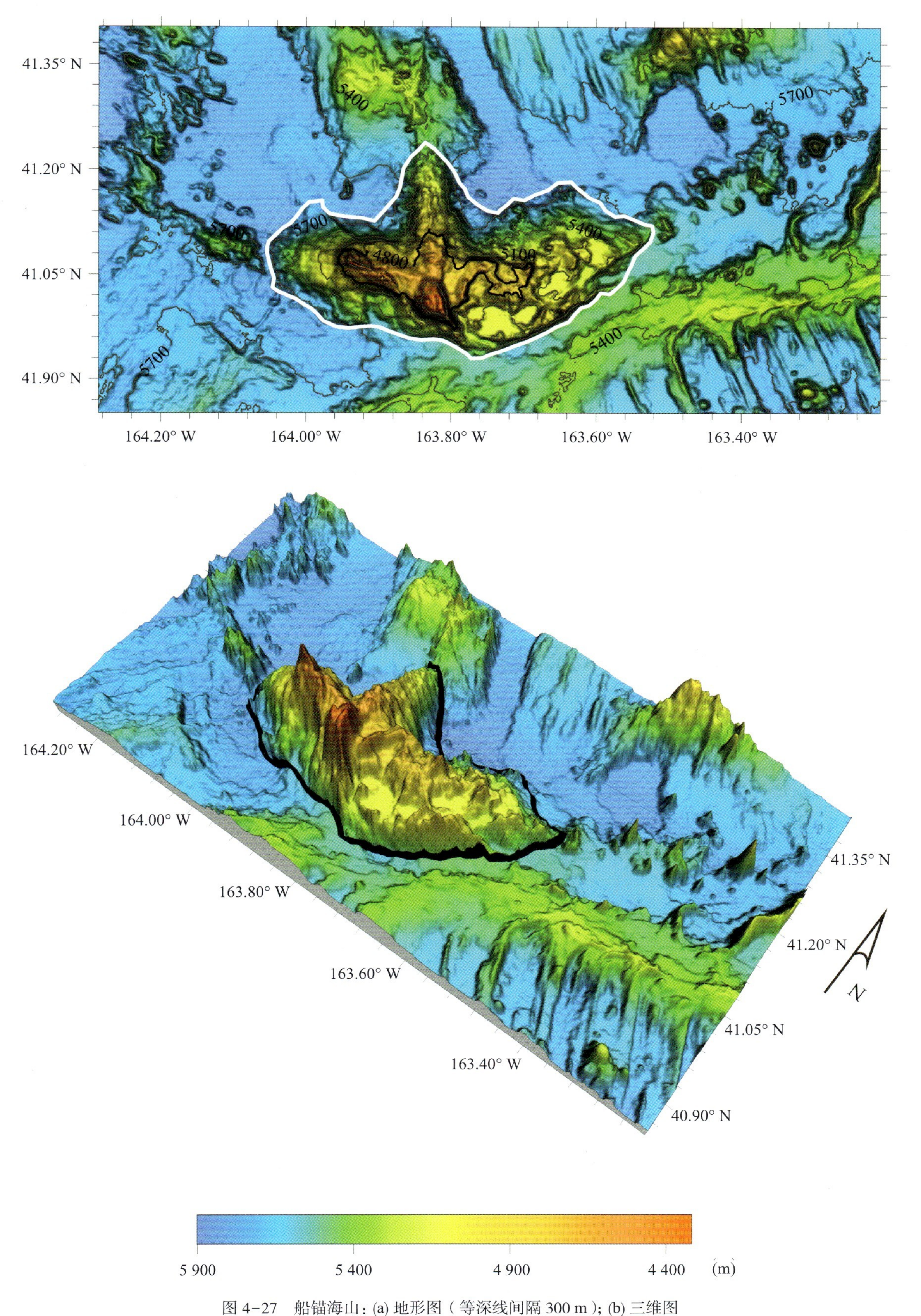

图 4-27 船锚海山：(a) 地形图（等深线间隔 300 m）；(b) 三维图

Fig.4-27 Chuanmao Seamount: (a) Bathymetric map (Contours are in 300 m); (b) 3-D topographic map

4.2.27 瑶琴海丘 Yaoqin Hill

<table>
<tr><td>中文名称
Chinese Name</td><td colspan="3">Yáoqín Hǎiqiū
瑶琴海丘</td></tr>
<tr><td>英文名称
English Name</td><td colspan="3">Yaoqin Hill</td></tr>
<tr><td>地理区域 / Location</td><td colspan="3">东北太平洋 The Northeast Pacific Ocean</td></tr>
<tr><td rowspan="2">特征点坐标
Coordinate</td><td rowspan="2">40°52.6′N 162°00.9′W</td><td>长度 / Length</td><td>14 km</td></tr>
<tr><td>宽度 / Width</td><td>10 km</td></tr>
<tr><td>水深 / Depth</td><td>4 655~5 603 m</td><td>高差 / Total Relief</td><td>948 m</td></tr>
<tr><td>发现情况
Discovery Facts</td><td colspan="3">此海丘于 2020 年“向阳红 01”船在执行航次调查时发现。
The hill was discovered in 2020 during the survey cruise carried out onboard the Chinese R/V Xiang Yang Hong 01.</td></tr>
<tr><td>地形特征
Feature Description</td><td colspan="3">瑶琴海丘发育于太平洋东北部，位于音乐家海山群以北 1 080 km，长和宽分别为 14 km 和 10 km。该海丘顶部发育小规模凸起，水深 4 655 m，两侧陡峭，底部水深 5 603 m，最大高差约 948 m。
Located in the northeast of the Pacific Ocean, the hill is 1,080 km north of the Musicians Seamounts. It is 14 km long and 10 km wide. There are a few protrusions at the top of the hill, and the two flanks of the hill are steep. It is 4,655 m deep at the top and 5,603 m deep at the foot, with a maximum total relief of about 948 m.</td></tr>
<tr><td>专名释义
Reason for Choice of Name</td><td colspan="3">该区域沿用中国乐器相关命名体系群组化方法命名。瑶琴，又称古琴，是中国传统拨弦乐器，音域宽广，音色深沉，余音悠远。故以瑶琴命名。
The features in this area are all named following the Chinese musical instrument naming system. The Yaoqin, also known as Guqin, is a traditional Chinese plucked string instrument with a wide range of tones, deep timbre and a lingering sound. Hence, the feature is named Yaoqin Hill.</td></tr>
</table>

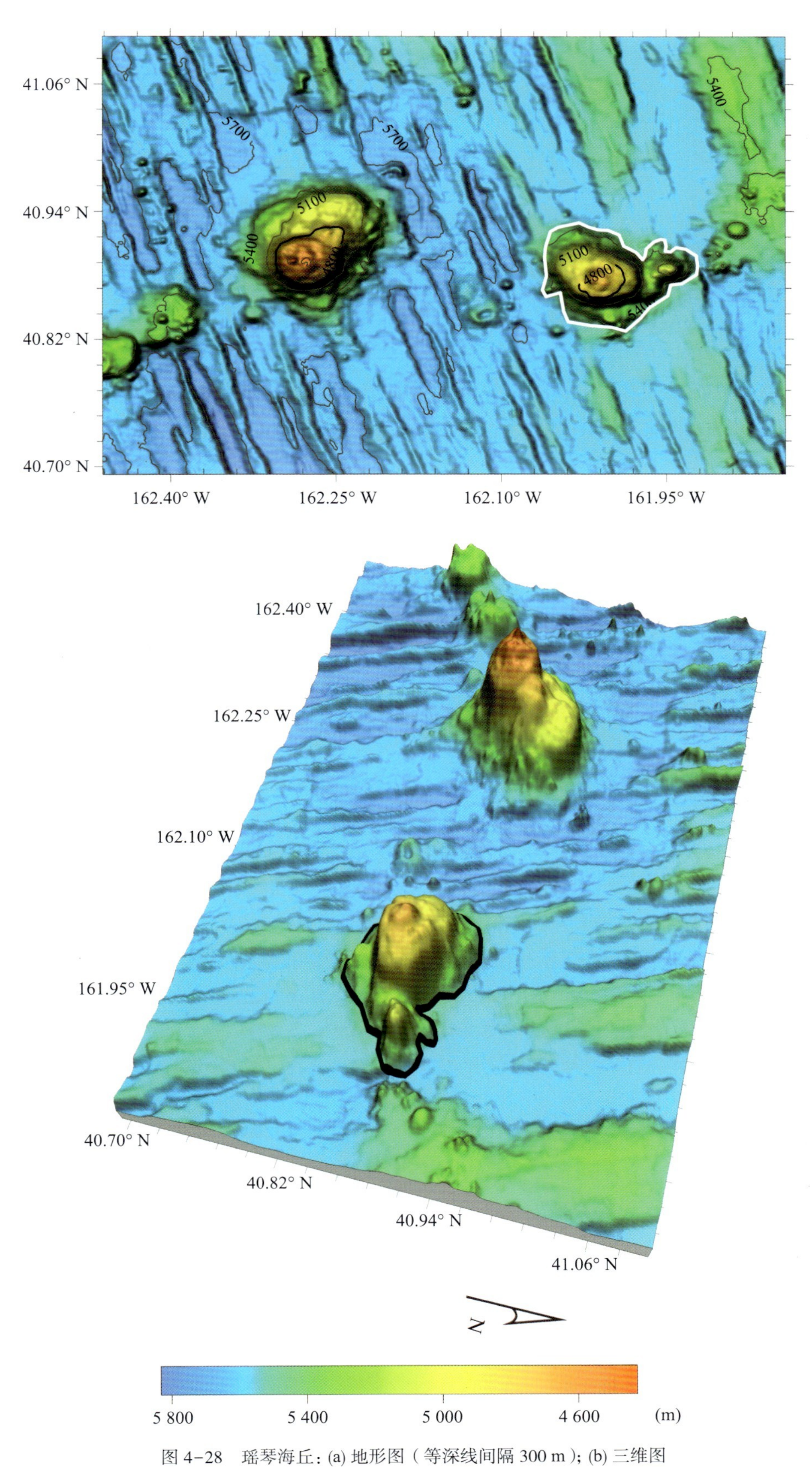

图 4-28 瑶琴海丘：(a) 地形图（等深线间隔 300 m）；(b) 三维图
Fig.4-28 Yaoqin Hill: (a) Bathymetric map (Contours are in 300 m); (b) 3-D topographic map

4.2.28 阮咸圆丘 Ruanxian Knoll

<table>
<tr><td>中文名称
Chinese Name</td><td colspan="3">Ruǎnxián Yuánqiū
阮咸圆丘</td></tr>
<tr><td>英文名称
English Name</td><td colspan="3">Ruanxian Knoll</td></tr>
<tr><td>地理区域 / Location</td><td colspan="3">东北太平洋 The Northeast Pacific Ocean</td></tr>
<tr><td rowspan="2">特征点坐标
Coordinate</td><td rowspan="2">39°31.4′N 160°55.7′W</td><td>长度 / Length</td><td>11 km</td></tr>
<tr><td>宽度 / Width</td><td>10 km</td></tr>
<tr><td>水深 / Depth</td><td>4 837~5 633 m</td><td>高差 / Total Relief</td><td>796 m</td></tr>
<tr><td>发现情况
Discovery Facts</td><td colspan="3">此圆丘于 2019 年“向阳红 06”船在执行航次调查时发现。
The knoll was discovered in 2019 during the survey cruise carried out onboard the Chinese R/V Xiang Yang Hong 06.</td></tr>
<tr><td>地形特征
Feature Description</td><td colspan="3">阮咸圆丘发育于太平洋东北部，位于音乐家海山群以北930 km，整体呈圆形，长和宽分别为 11 km 和 10 km。该圆丘顶部相对平坦，中部发育小型凸起，最高处水深 4 837 m，山麓水深 5 633 m，最大高差约 796 m。
Located in the northeast of the Pacific Ocean, the knoll is 930 km north of the Musicians Seamounts. Being rounded in shape, the knoll is 11 km long and 10 km wide. With a flat top, the knoll has small protrusions in the middle. It is 4,837 m deep at the highest point and 5,633 m deep at the piedmont, with a maximum total relief of about 796 m.</td></tr>
<tr><td>专名释义
Reason for Choice of Name</td><td colspan="3">该区域沿用中国乐器相关命名体系群组化方法命名。阮咸，我国传统弹拨乐器。相传西晋阮咸善弹此乐器，因而得名。该圆丘主体浑圆，形似阮咸圆形共鸣箱，故以阮咸命名。
The features in this sea area are all named following the Chinese musical instrument naming system. The Ruanxian is a traditional plucked string instrument of China. It is said that there was a famous musician named Ruanxian in the Western Jin Dynasty who was good at playing this instrument, hence the instrument was named Ruanxian. The main body of the knoll is rounded, and shaped like a round resonator of Ruanxian. Hence, the knoll is named Ruanxian Knoll.</td></tr>
</table>

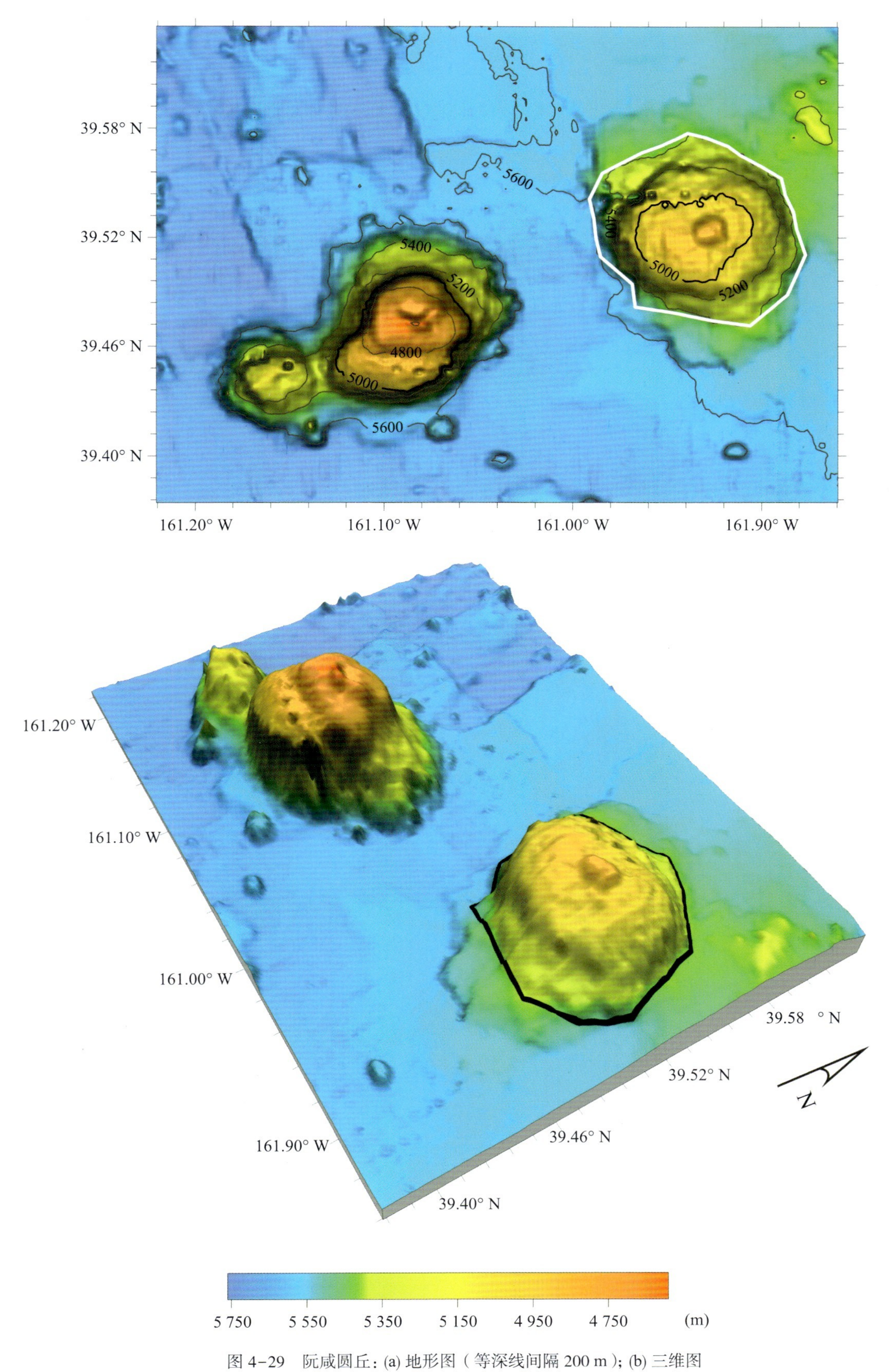

图 4-29 阮咸圆丘：(a) 地形图（等深线间隔 200 m）；(b) 三维图

Fig.4-29 Ruanxian Knoll: (a) Bathymetric map (Contours are in 200 m); (b) 3-D topographic map

5

东印度洋海底地理实体

Chapter V

Undersea Features in the Eastern Indian Ocean

5.1 地形地貌概况

印度洋是世界第三大洋，面积约 74.1 × 10^6 km^2，南北长约 9 600 km，东西宽约 7 800 km。北部以伊朗、巴基斯坦、印度和孟加拉国为界，东部以马来半岛、印度尼西亚巽他群岛和澳大利亚为界，西南与大西洋相接，东部及东南部与太平洋相接，平均深度约 3 890 m，最深处为 8 047 m。

在印度洋海底中部分布着大型延伸的中央海岭，把印度洋分为东印度洋、西印度洋和南印度洋三大海域。东印度洋被东印度洋海岭（东经九十度海岭）分隔为中印度洋海盆、沃顿海盆、科科斯海盆等大型海底地理实体。这些海盆均比较广阔，海水较深。

东印度洋海域最为显著的地形是东经九十度海岭，其位于印度洋东部 90°E 位置，沿经线呈线形延伸，是世界上最长直的海岭。东经九十度海岭北起 10°N，南至 32°S，长度约 4 000 km，宽度 100~200 km，相对高差约 2 000 m。基部水深约 4 000 m，顶部平均水深约 2 000 m，地形比 较平缓。该海岭北端伸入孟加拉湾，淹没于巨厚的孟加拉冲积扇之下；南端与布罗肯海脊（Broken Ridge）相连。海岭以南为东南印度洋脊，两者最近距离约 900 km；东西两侧分别为广袤而平坦的沃顿海盆和中印度洋海盆，平均水深约 5 000 m。

Section 5.1 Overview of the topography

The Indian Ocean is the third largest ocean in the world, with an area of approximately 74.1 × 10^6 km^2, approximately 9,600 km long from north to south, and about 7,800 km wide from east to west. Its northern part is bounded by Iran, Pakistan, India, and Bangladesh, while the eastern part is bounded by the Malay Peninsula, Indonesia's Sunda Islands, and Australia. It is bordered by the Atlantic Ocean to the southwest and the Pacific Ocean to the east and southeast, with an average depth of about 3,890 m and a maximum depth of 8,047 m.

In the central part of the Indian Ocean, there is a large extended central ridge that divides the Indian Ocean into three major sea areas: the East Indian Ocean, the West Indian Ocean, and the South Indian Ocean. The East Indian Ocean is divided by the East Indian Ocean Ridge at 90°E (the Ninety East

Ridge) into the Central Indian Ocean Basin, Wharton Basin, and Cocos Basin. These basins are relatively vast and have a relatively large water depth.

The Ninety East Ridge is located in the Eastern Indian Ocean at 90°E and extends linearly along the longitude, making it the longest and straightest ridge in the world. It starts from 10°N and ends at 32°S, about 4,000 km length and 100~200 km width, with a relative height difference of about 2,000 m. The depth at the base is about 4,000 m; the average water depth at the top is about 2,000 m, and the relief is relatively gentle. The northern end of the Ridge extends into the Bay of Bengal and is submerged under the thick Bengal alluvial fan; the southern end is connected to the Broken Ridge. To the south of the Ninety East Ridge is the Southeast Indian Ridge. The closest distance between these two ridges is about 900 km, on both sides are the vast and flat Walton Basin and the Central Indian Ocean Basin, with an average water depth of about 5,000 m.

5.2 东印度洋地理实体命名

印度洋是古代海上丝绸之路的重要通道，它见证了中国茶叶、瓷器和丝绸穿越其波澜壮阔的水域，输往遥远的欧洲和沿线国家。这条古老的航道将东方的瑰宝和西方的向往紧密联系在一起，以绵延不绝的航迹串联起遥远的文明与梦想。因此，东印度洋的海底地理实体命名，主要采用中国茶文化群组氏命名方法，以象征中国通过印度洋与西方各国友好的贸易和文化往来，同时辅以象形、东印度洋生物物种等进行指位性与象征性命名。

新命名的海底地理实体主要分布在东经九十度海岭海域，沿海岭东西两侧分布。新命名的海底地理实体共计 15 个，其中海底水道 2 个，为九十度东海底水道、岭西南海底水道；海山和海山群 2 个，为红袍海山、普洱海山群；海丘和海丘群 6 个，为桨躄海丘、玉露海丘、猴魁海丘、泉茗海丘、银梭海丘、雀舌海丘群；圆丘 1 个，为金沱圆丘；海脊 4 个，为毛尖海脊、茗眉海脊、仙毫海脊和银针海脊。

Section 5.2 Undersea features in the Eastern Indian Ocean

The Indian Ocean was an important gateway for the ancient maritime Silk Road, witnessing the Chinese tea, porcelain, and silk across its vast expanse to distant lands including Europe. This ancient maritime route connected the treasures of the East with the aspirations of the West, weaving together distant civilizations and dreams through the Indian Ocean. Therefore, the undersea features in the East Indian Ocean are named mainly following the Chinese tea culture naming system to symbolize China's friendly trade and cultural exchanges with Western countries through the Indian Ocean, while supplemented by pictograms and East Indian Ocean biological species for positional and symbolic names.

These undersea features are mainly distributed on the east and west sides of the Ninety East Ridge in the East Indian Ocean. A total of fifteen undersea features newly discovered on both sides of the Ninety East Ridge in the Eastern Indian Ocean have been named, including two sea channels: Jiushidudong Sea Channel and Lingxinan Sea Channel; one seamount and one relatively gathering seamounts (seamount group): Hongpao Seamount and Pu'er Seamounts; six hills including one relatively gathering hills (hill group): Jiangbi Hill, Yulu Hill, Houkui Hill, Quanming Hill, Yinsuo Hill, and Queshe Hills; one knoll: Jintuo Knoll ; four ridges: Maojian Ridge, Mingmei Ridge, Xianhao Ridge and Yinzhen Ridge.

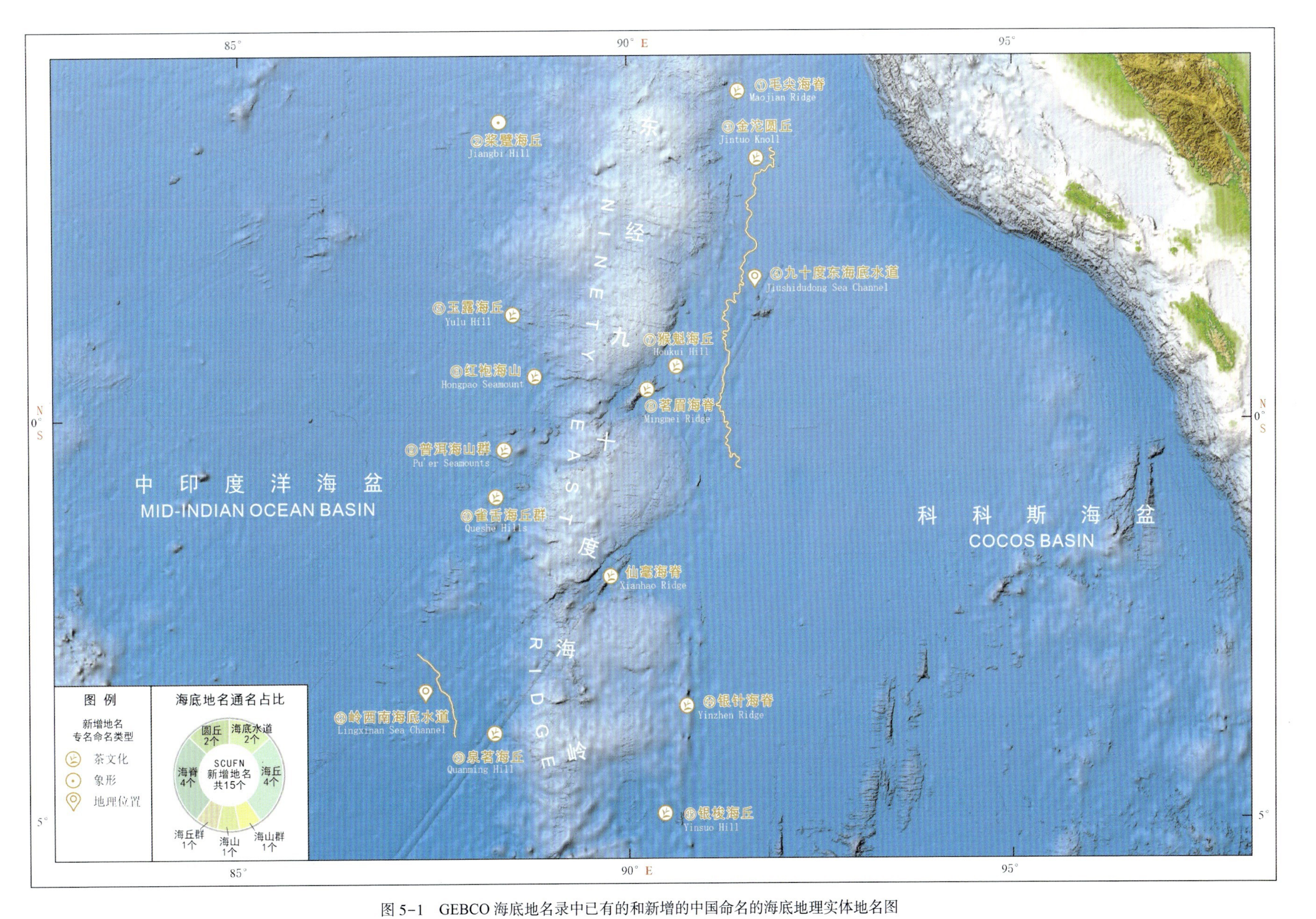

图 5-1 GEBCO 海底地名录中已有的和新增的中国命名的海底地理实体地名图

Fig.5-1 Map of the undersea feature names proposed by China and already included in GEBCO Gazetteer and these newly proposed by China

1	毛尖海脊	Maojian Ridge	04°06.1′N 91°27.3′E
2	桨甓海丘	Jiangbi Hill	03°43.7′N 88°21.5′E
3	金沱圆丘	Jintuo Knoll	03°15.9′N 91°41.6′E
4	九十度东海底水道	Jiushidudong Sea Channel	00°29.3′N 91°14.9′E
5	玉露海丘	Yulu Hill	01°18.7′N 88°30.8′E
6	红袍海山	Hongpao Seamount	00°31.9′N 88°48.0′E
7	猴魁海丘	Houkui Hill	00°39.7′N 90°37.5′E
8	茗眉海脊	Mingmei Ridge	00°22.0′N 90°14.8′E
9	普洱海山群	Pu’er Seamounts	00°21.6′S 88°37.5′E 00°22.7′S 88°23.8′E
10	雀舌海丘群	Queshe Hills	00°57.6′S 88°18.9′E 00°57.6′S 88°10.0′E 00°58.4′S 88°17.4′E
11	仙毫海脊	Xianhao Ridge	01°58.1′S 89°46.1′E
12	岭西南海底水道	Lingxinan Sea Channel	03°20.6′S 87°33.4′E
13	泉茗海丘	Quanming Hill	03°55.7′S 88°16.1′E
14	银针海脊	Yinzhen Ridge	03°35.5′S 90°45.5′E
15	银梭海丘	Yinsuo Hill	04°56.1′N 90°28.2′E

5.2.1 毛尖海脊 Maojian Ridge

中文名称 Chinese Name	Máojiān Hǎijǐ 毛尖海脊		
英文名称 English Name	Maojian Ridge		
地理区域 / Location	东印度洋 The East Indian Ocean		
特征点坐标 Coordinate	04°06.1′N 91°27.3′E	**长度 / Length**	38 km
		宽度 / Width	8 km
水深 / Depth	3 349~4 032 m	**高差 / Total Relief**	683 m
发现情况 Discovery Facts	此海脊于 2019 年“向阳红 01”船在执行航次调查时发现。 The ridge was discovered in 2019 during the survey cruise carried out onboard the Chinese R/V *Xiang Yang Hong 01*.		
地形特征 Feature Description	毛尖海脊位于东印度洋，紧邻东经九十度海岭东部，呈 N-S 向条状延伸，北部狭长，中部和南部略宽，南北长 38 km，东西宽 8 km。该山脊顶部高耸，东侧呈断崖地势。最高处位于海脊中部，水深约 3 349 m。 Located in the East Indian Ocean, the ridge is adjacent to the east of the Ninety East Ridge, and extends in the N-S direction like a stripe. It is long and narrow in the north, slightly wider in the middle and south, with a length of 38 km from south to north, and a width of 8 km from west to east. With a towering peak, the ridge takes the shape of a precipice in the east. The highest point is located in the middle of the ridge, with a depth of about 3,349 m.		
专名释义 Reason for Choice of Name	印度洋是古代海上丝绸之路的重要通道，中国将茶叶、瓷器和丝绸穿越印度洋运往欧洲等地的国家，将东西方联系在一起。该区域采用中国茶文化群组化方法命名，以象征中国通过印度洋与西方各国友好的贸易和文化往来。毛尖茶，是中国绿茶的一个子品种，产于中国河南信阳一带。故以毛尖命名。 The Indian Ocean was an important channel of the ancient Maritime Silk Road. China transported tea, porcelain and silk across the Indian Ocean to countries including those in Europe, which strengthened the ties between the East and the West. The features in this sea area are all named following the Chinese tea culture naming system to symbolize the friendly trade and cultural exchanges between China and Western countries through the Indian Ocean. Maojian tea is a kind of the Chinese green tea produced in Xinyang, Henan Province of China. Hence, the feature is named Maojian Ridge.		

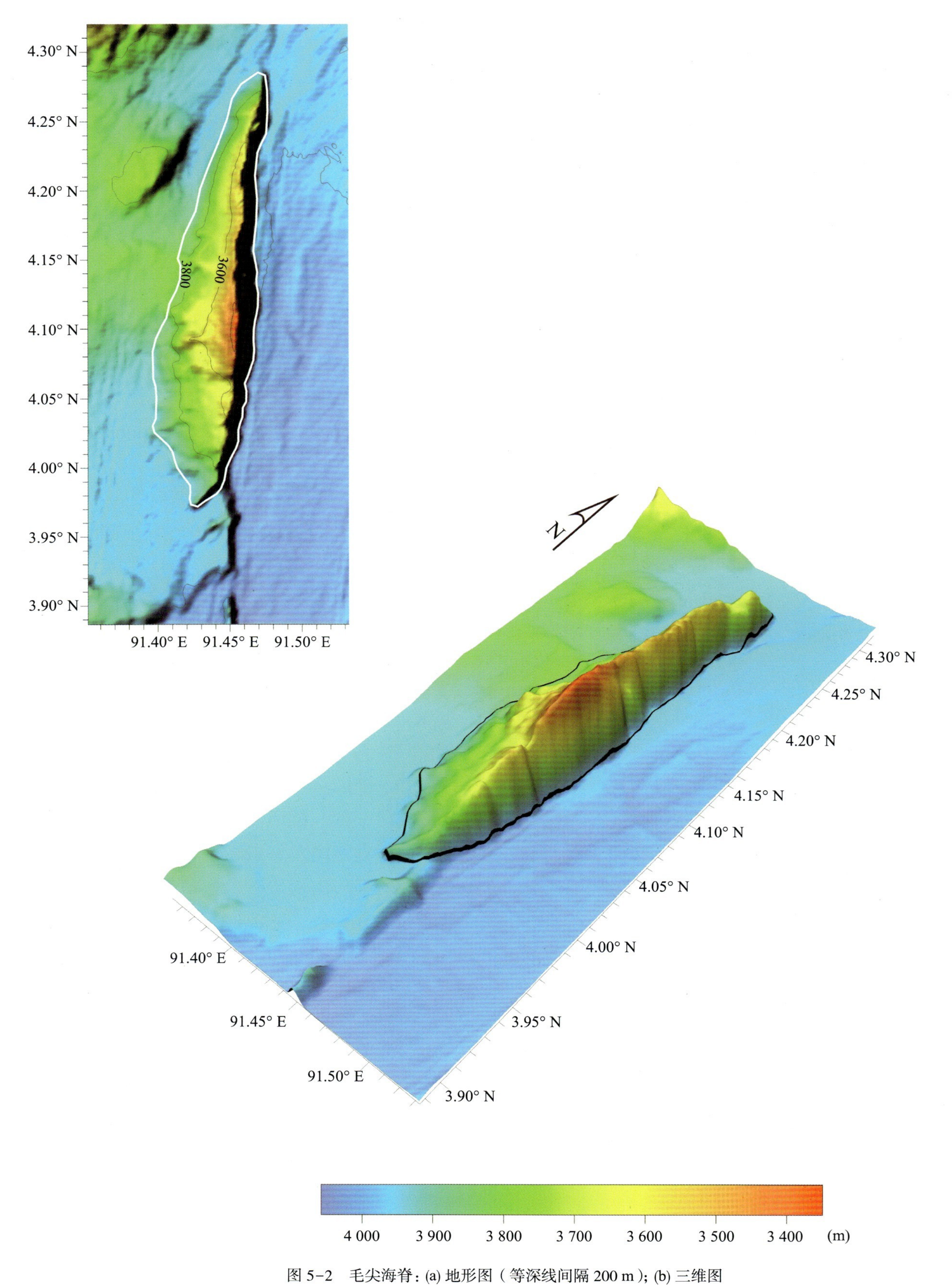

图 5-2　毛尖海脊：(a) 地形图（等深线间隔 200 m）；(b) 三维图

Fig.5-2　Maojian Ridge:(a) Bathymetric map (Contours are in 200 m); (b) 3-D topographic map

5.2.2 桨鳖海丘 Jiangbi Hill

<table>
<tr><td>中文名称
Chinese Name</td><td colspan="3">Jiǎngbì Hǎiqiū
桨鳖海丘</td></tr>
<tr><td>英文名称
English Name</td><td colspan="3">Jiangbi Hill</td></tr>
<tr><td>地理区域 / Location</td><td colspan="3">东印度洋 The East Indian Ocean</td></tr>
<tr><td rowspan="2">特征点坐标
Coordinate</td><td rowspan="2">03°43.7′N 88°21.5′E</td><td>长度 / Length</td><td>20.9 km</td></tr>
<tr><td>宽度 / Width</td><td>6.8 km</td></tr>
<tr><td>水深 / Depth</td><td>3 766~4 075 m</td><td>高差 / Total Relief</td><td>309 m</td></tr>
<tr><td>发现情况
Discovery Facts</td><td colspan="3">此海丘于 2019 年“向阳红 01”船在执行航次调查时发现。
The hill was discovered in 2019 during the survey cruise carried out onboard the Chinese R/V Xiang Yang Hong 01.</td></tr>
<tr><td>地形特征
Feature Description</td><td colspan="3">桨鳖海丘发育于东印度洋东经九十度海岭以西 30 km，呈 NE-SW 向弯曲延伸，该海丘总长约 20.9 km，宽约 6.8 km。水深范围 3 766~4 075 m，最大高差约 309 m，位于海丘东北部。
The hill is 30 km west of the Ninety East Ridge in the East Indian Ocean, and extends in the NE-SW direction, with a total length of about 20.9 km and a width of about 6.8 km. Its depth varies from 3,766 m to 4,075 m, with a maximum total relief of about 309 m in the location northeast of the hill.</td></tr>
<tr><td>专名释义
Reason for Choice of Name</td><td colspan="3">桨鳖是一种栖息于印度洋、太平洋等海域的鱼类。该海丘西南部凸出如鱼首，东北部山体散开如鱼尾，东南部凸出似鱼鳍，俯视整体犹如一尾游向大洋深处的桨鳖鱼，故以桨鳖命名。
Jiangbi is the Chinese name for a kind of fish inhabiting in the sea areas of the Indian Ocean, the Pacific Ocean, and so on. As its southwestern part protruding like the fish head, the northeastern part spreading out like the fish tail, and the southeastern part protruding like the fish fin, the hill looks like a Jiangbi swimming deep into the ocean viewed from above. Hence, the feature is named Jiangbi Hill.</td></tr>
</table>

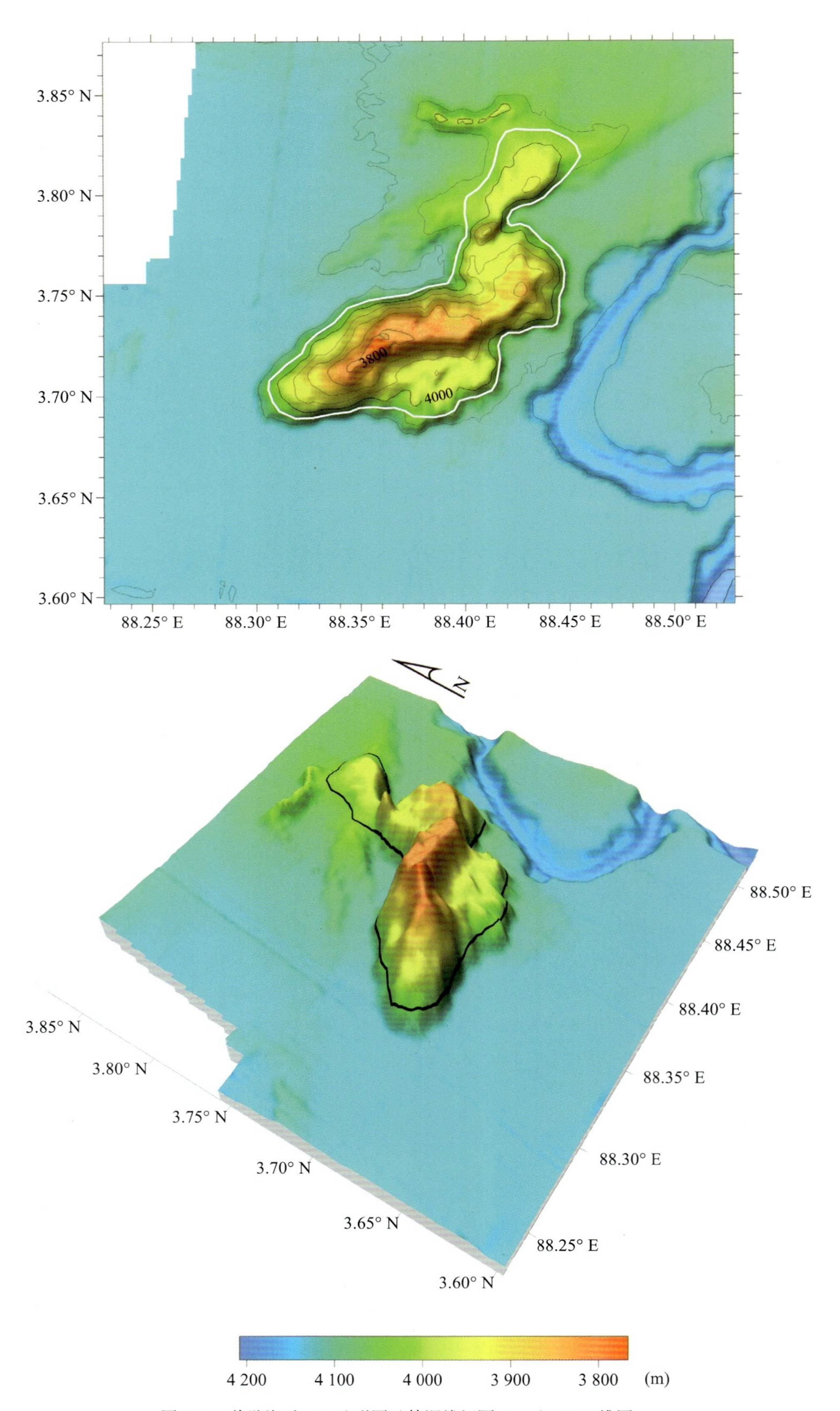

图 5-3　桨甓海丘：(a) 地形图（等深线间隔 50 m）；(b) 三维图

Fig.5-3 Jiangbi Hill: (a) Bathymetric map (Contours are in 50 m); (b) 3-D topographic map

5.2.3 金沱圆丘 Jintuo Knoll

<table>
<tr><td>中文名称
Chinese Name</td><td colspan="3">Jīntuó Yuánqiū
金沱圆丘</td></tr>
<tr><td>英文名称
English Name</td><td colspan="3">Jintuo Knoll</td></tr>
<tr><td>地理区域 / Location</td><td colspan="3">东印度洋 The East Indian Ocean</td></tr>
<tr><td rowspan="2">特征点坐标
Coordinate</td><td rowspan="2">03°15.9′N 91°41.6′E</td><td>长度 / Length</td><td>6.1 km</td></tr>
<tr><td>宽度 / Width</td><td>5 km</td></tr>
<tr><td>水深 / Depth</td><td>3 471~4 126 m</td><td>高差 / Total Relief</td><td>655 m</td></tr>
<tr><td>发现情况
Discovery Facts</td><td colspan="3">此海丘于 2019 年“向阳红 01”船在执行航次调查时发现。
The knoll was discovered in 2019 during the survey cruise carried out onboard the Chinese R/V Xiang Yang Hong 01.</td></tr>
<tr><td>地形特征
Feature Description</td><td colspan="3">金沱圆丘孤立于东印度洋东经九十度海岭以东 50 km，整体呈等维展布。该海丘东侧发育一 S-N 走向带状凹陷，将圆丘分为东西两个部分，西部为圆丘主体，最高处水深 3 471 m，东部规模较小，地势平缓。
Being 50 km east of the Ninety East Ridge in the East Indian Ocean, the knoll is equidimensional in plan. In the east, there is a band-shaped depression extending in the S-N direction. The depression divides the knoll into two parts, of which the western part serves as the main body with a depth of 3,471 m at the highest point, and the eastern part is smaller and flat.</td></tr>
<tr><td>专名释义
Reason for Choice of Name</td><td colspan="3">该区域采用中国茶文化群组化方法命名，以象征中国通过印度洋与西方各国友好的贸易和文化往来。金沱，是我国云南的一种普洱茶。此圆丘形体浑圆，形似普洱小金沱，故名。
The features in this sea area are all named following the Chinese tea culture naming system to symbolize the friendly trade and cultural exchanges between China and Western countries through the Indian Ocean. Jintuo is the name of a kind of Pu’er tea produced in Yunnan Province, China. The rounded knoll looks like a Pu’er Jintuo tea leaf. Hence, the knoll is named Jintuo Knoll.</td></tr>
</table>

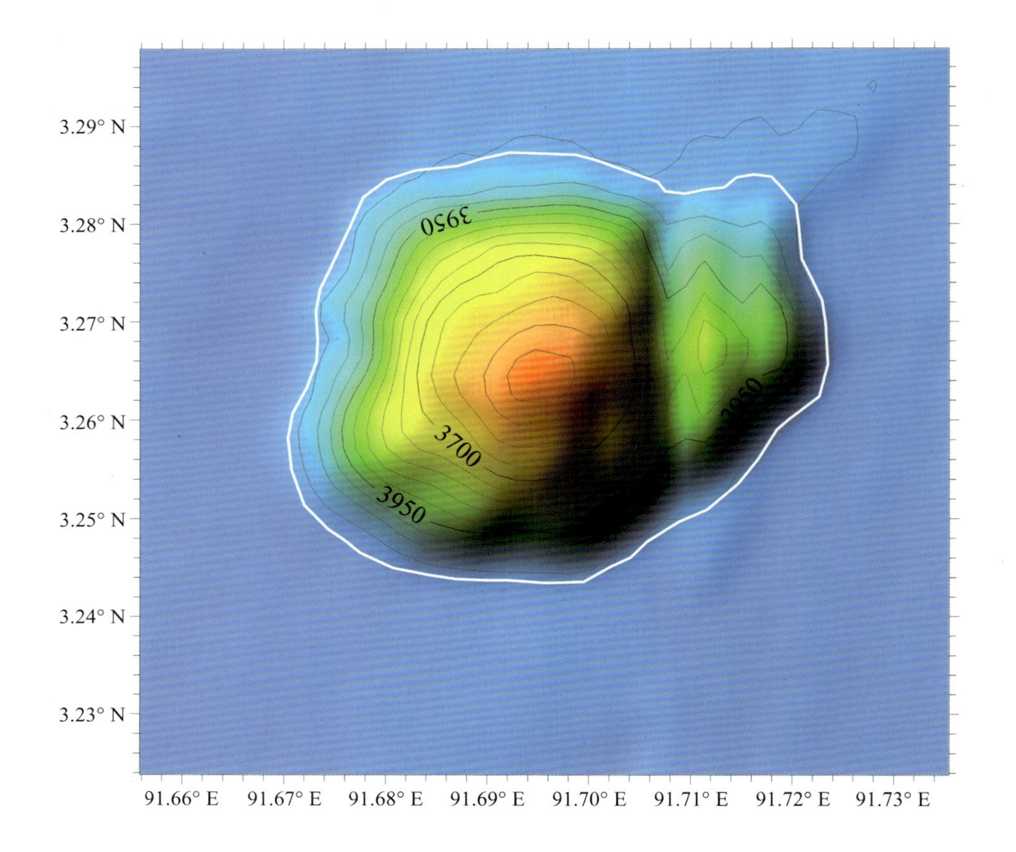

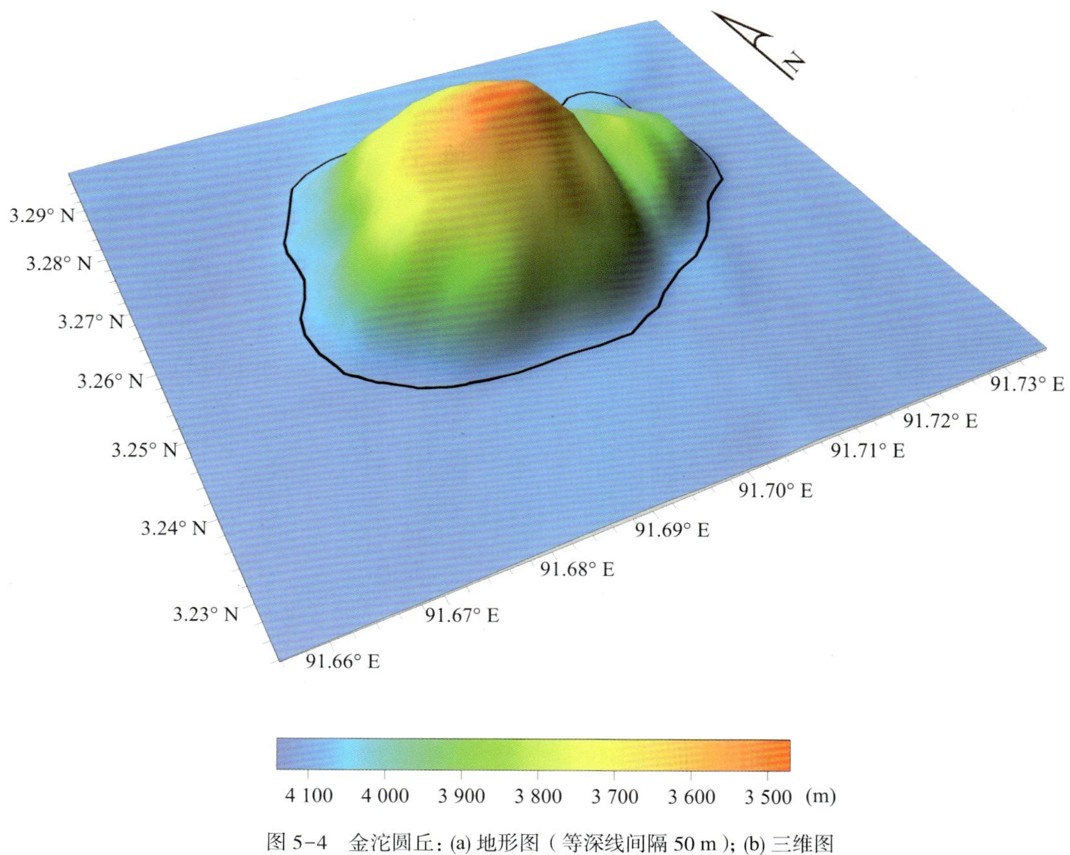

图 5-4　金沱圆丘：(a) 地形图（等深线间隔 50 m）；(b) 三维图

Fig.5-4　Jintuo Knoll: (a) Bathymetric map (Contours are in 50 m);(b)3-D topographic map

5.2.4 九十度东海底水道 Jiushidudong Sea Channel

中文名称 Chinese Name	Jiǔshídùdōng Hǎidǐshuǐdào 九十度东海底水道		
英文名称 English Name	Jiushidudong Sea Channel		
地理区域 / Location	东印度洋 The East Indian Ocean		
特征点坐标 Coordinate	00°29.3′N 91°14.9′E	长度 / Length	215.19 km
		宽度 / Width	3 km
水深 / Depth	4 178~4 635 m	高差 / Total Relief	457 m
发现情况 Discovery Facts	此水道于 2019 年“向阳红 01”船在执行航次调查时发现。 The sea channel was discovered in 2019 during the survey cruise carried out onboard the Chinese R/V *Xiang Yang Hong 01*.		
地形特征 Feature Description	九十度东海底水道发育于东印度洋东经九十度海岭以东 30 km，呈 N-S 向蜿蜒延伸，南北长超过 200 km，东西宽 3 km，平均水深约 4 364 m。 Located 30 km east of the Ninety East Ridge in the East Indian Ocean, the sea channel winds in the N-S direction, and it is more than 200 km long from south to north, and 3 km wide from west to east, with an average depth of about 4,364 m.		
专名释义 Reason for Choice of Name	该海底水道位于东经九十度海岭以东，故以九十度东命名。 The sea channel is located to the east of the Ninety East Ridge. Hence, the sea channel is named Jiushidudong Sea Channel.		

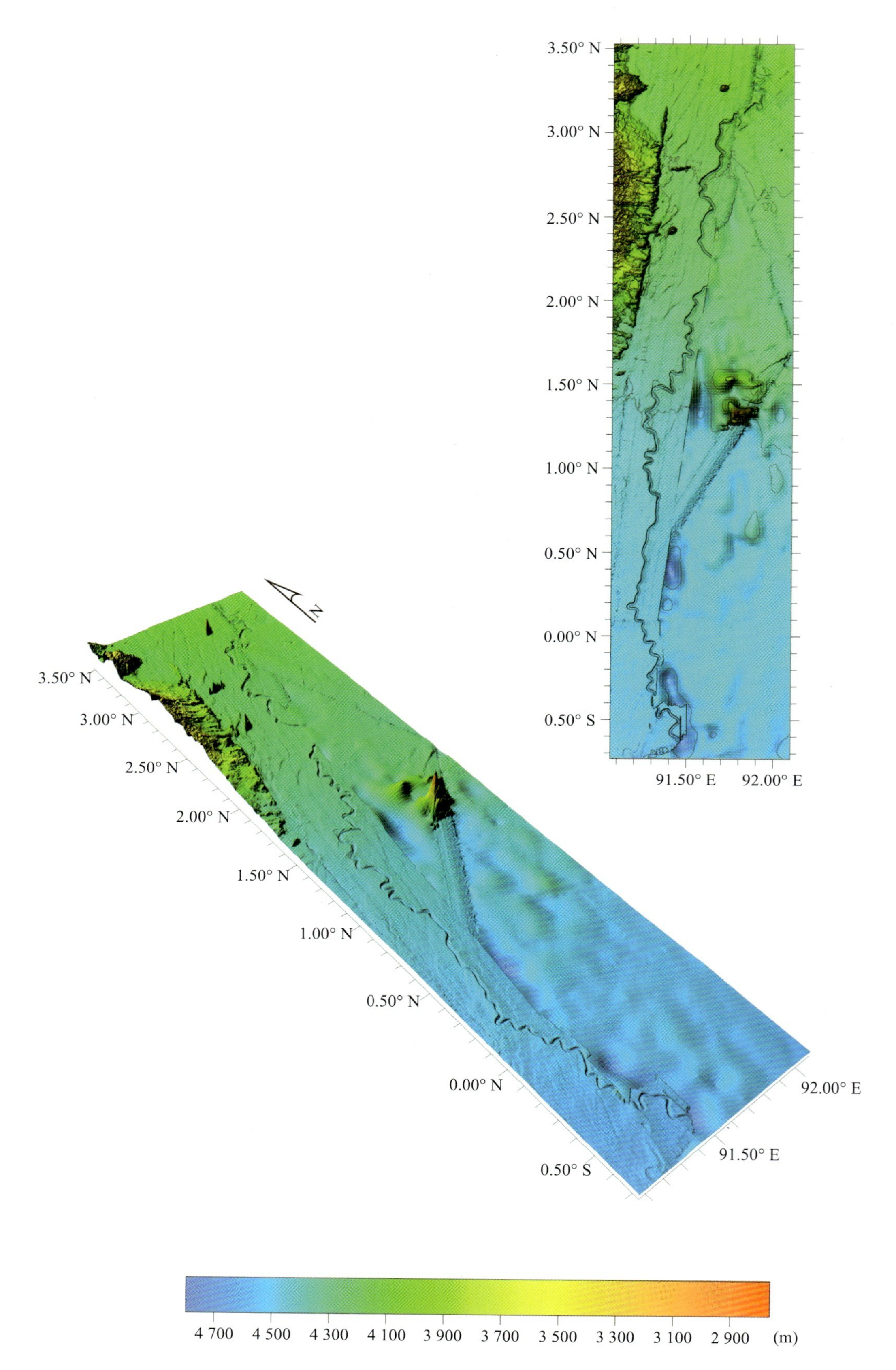

图 5-5　九十度东海底水道：(a) 地形图（等深线间隔 200 m）；(b) 三维图

Fig.5-5　Jiushidudong Sea Channel: (a) Bathymetric map (Contours are in 200 m); (b) 3-D topographic map

5.2.5 玉露海丘 Yulu Hill

中文名称 Chinese Name	Yùlù Hǎiqiū 玉露海丘		
英文名称 English Name	Yulu Hill		
地理区域 / Location	东印度洋 The East Indian Ocean		
特征点坐标 Coordinate	01°18.7′N 88°30.8′E	长度 / Length 宽度 / Width	8.6 km 5.6 km
水深 / Depth	3 644~4 354 m	高差 / Total Relief	710 m
发现情况 Discovery Facts	此海丘于 2019 年“向阳红 01”船在执行航次调查时发现。 The hill was discovered in 2019 during the survey cruise carried out onboard the Chinese R/V *Xiang Yang Hong 01*.		
地形特征 Feature Description	玉露海丘位于东印度洋，紧邻东经九十度海岭西部，呈 NE-SW 走向，长约 8.6 km，宽约 5.6 km。该海丘最高处位于顶部东北部的尖状突起，水深 3 644 m，西南部较为平坦。 Located in the East Indian Ocean, the hill is adjacent to the west of the Ninety East Ridge. Extending in the NE-SW direction, it is about 8.6 km long and about 5.6 km wide. The highest point of the hill is the needle-shaped protrusion at the summit in the northeast, with a depth of 3,644 m. The southwest of the hill is relatively flat.		
专名释义 Reason for Choice of Name	该区域采用中国茶文化群组化方法命名，以象征中国通过印度洋与西方各国友好的贸易和文化往来，此海丘以玉露命名。玉露，是我国湖北恩施的一种特产茶。故以玉露命名。 The features in this sea area are all named following the Chinese tea culture naming system to symbolize the friendly trade and cultural exchanges between China and Western countries through the Indian Ocean. Yulu is the name of a specialty tea produced in Enshi, Hubei Province, China. Hence, the hill is named Yulu Hill.		

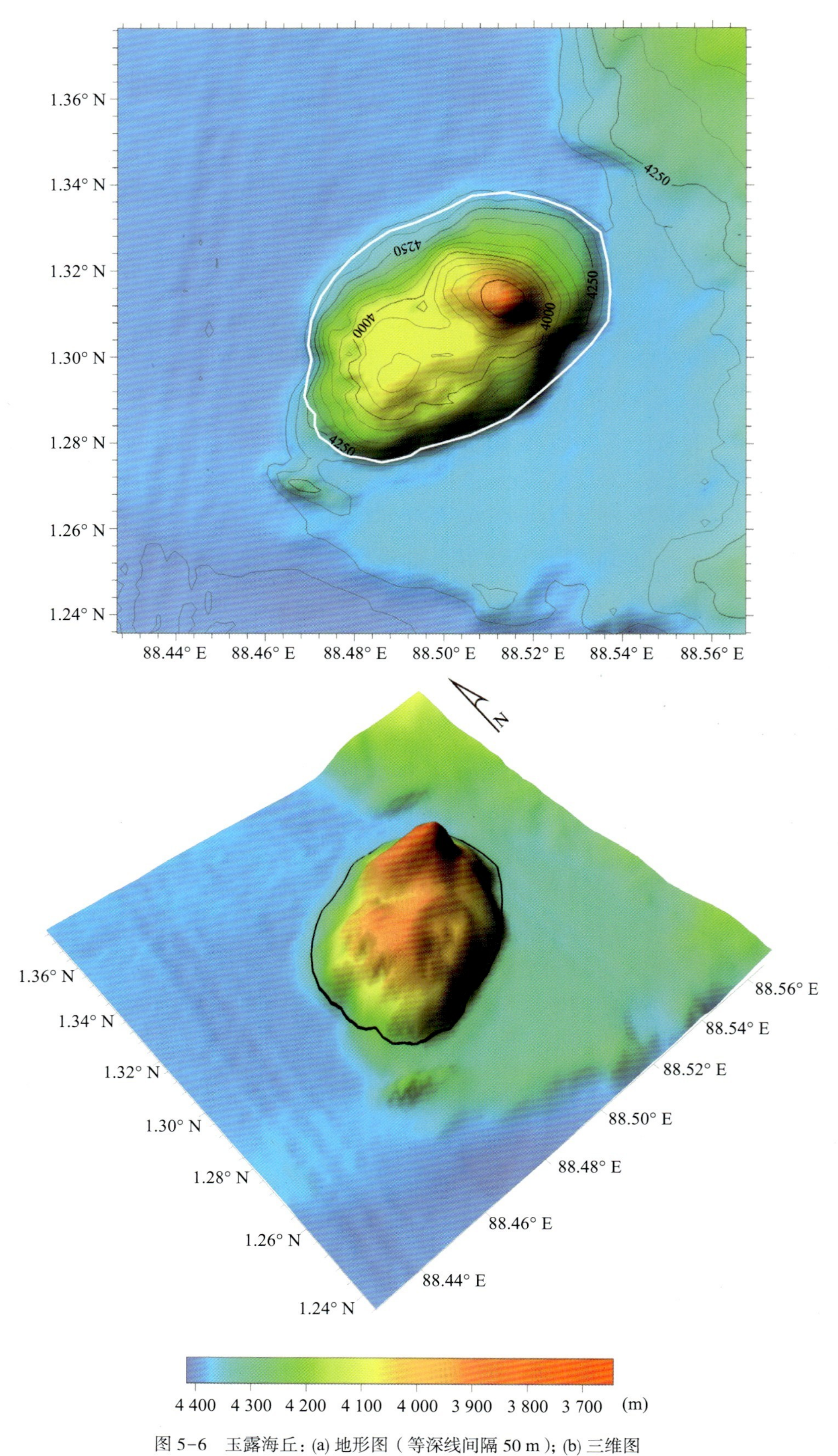

图 5-6 玉露海丘：(a) 地形图（等深线间隔 50 m）；(b) 三维图

Fig.5-6 Yulu Hill: (a) Bathymetric map (Contours are in 50 m); (b) 3-D topographic map

5.2.6 红袍海山 Hongpao Seamount

<table>
<tr><td>中文名称
Chinese Name</td><td colspan="3">Hóngpáo Hǎishān
红袍海山</td></tr>
<tr><td>英文名称
English Name</td><td colspan="3">Hongpao Seamount</td></tr>
<tr><td>地理区域 / Location</td><td colspan="3">东印度洋 The East Indian Ocean</td></tr>
<tr><td rowspan="2">特征点坐标
Coordinate</td><td rowspan="2">00°31.9′N 88°48.0′E</td><td>长度 / Length</td><td>21 km</td></tr>
<tr><td>宽度 / Width</td><td>16 km</td></tr>
<tr><td>水深 / Depth</td><td>3 134~4 468 m</td><td>高差 / Total Relief</td><td>1 334 m</td></tr>
<tr><td>发现情况
Discovery Facts</td><td colspan="3">此海山于 2019 年“向阳红 01”船在执行航次调查时发现。
The seamount was discovered in 2019 during the survey cruise carried out onboard the Chinese R/V Xiang Yang Hong 01.</td></tr>
<tr><td>地形特征
Feature Description</td><td colspan="3">红袍海山发育于东印度洋东经九十度海岭以西 16 km，主体呈 NE-SW 向延伸，长约 21 km，宽约 16 km。最高处位于山顶中部，水深 3 134 m。
The seamount is 16 km west of the Ninety East Ridge in the East Indian Ocean. Extending in the NE-SW direction, it is about 21 km long and about 16 km wide. The highest point is located in the middle of the summit, with a depth of 3,134 m.</td></tr>
<tr><td>专名释义
Reason for Choice of Name</td><td colspan="3">该区域采用中国茶文化群组化方法命名，以象征中国通过印度洋与西方各国友好的贸易和文化往来。此海山以红袍命名。大红袍，产于福建武夷山，属乌龙茶。
The features in this sea area are all named following the Chinese tea culture naming system to symbolize the friendly trade and cultural exchanges between China and Western countries through the Indian Ocean. The seamount is named Hongpao Seamount. Hongpao is the name of a kind of oolong tea produced in Mount Wuyi of Fujian Province, China.</td></tr>
</table>

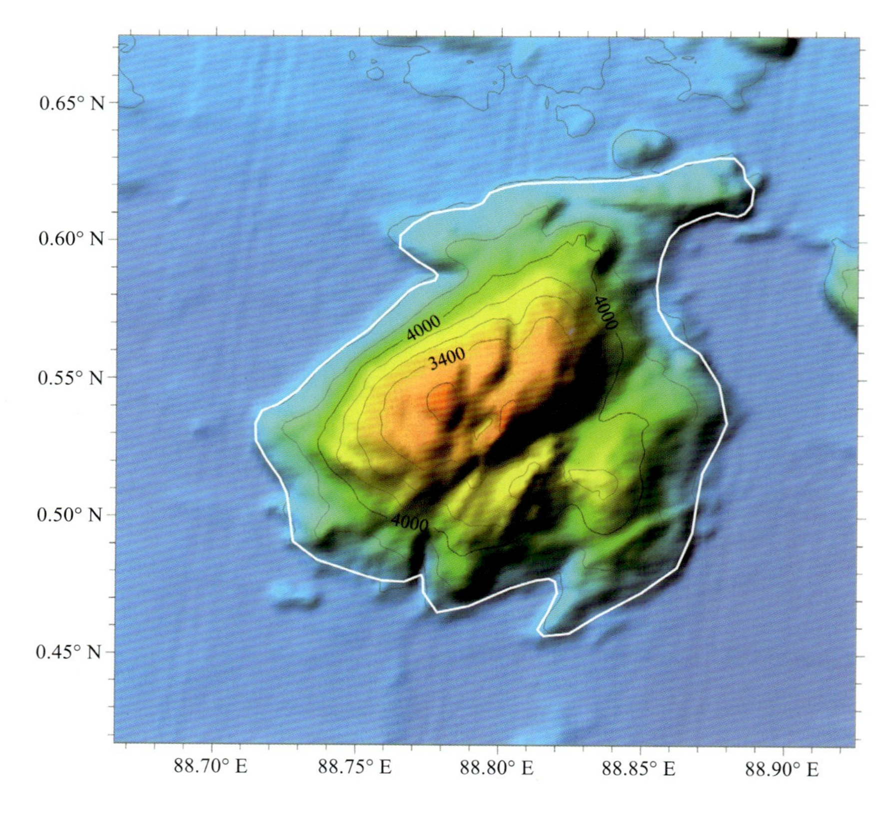

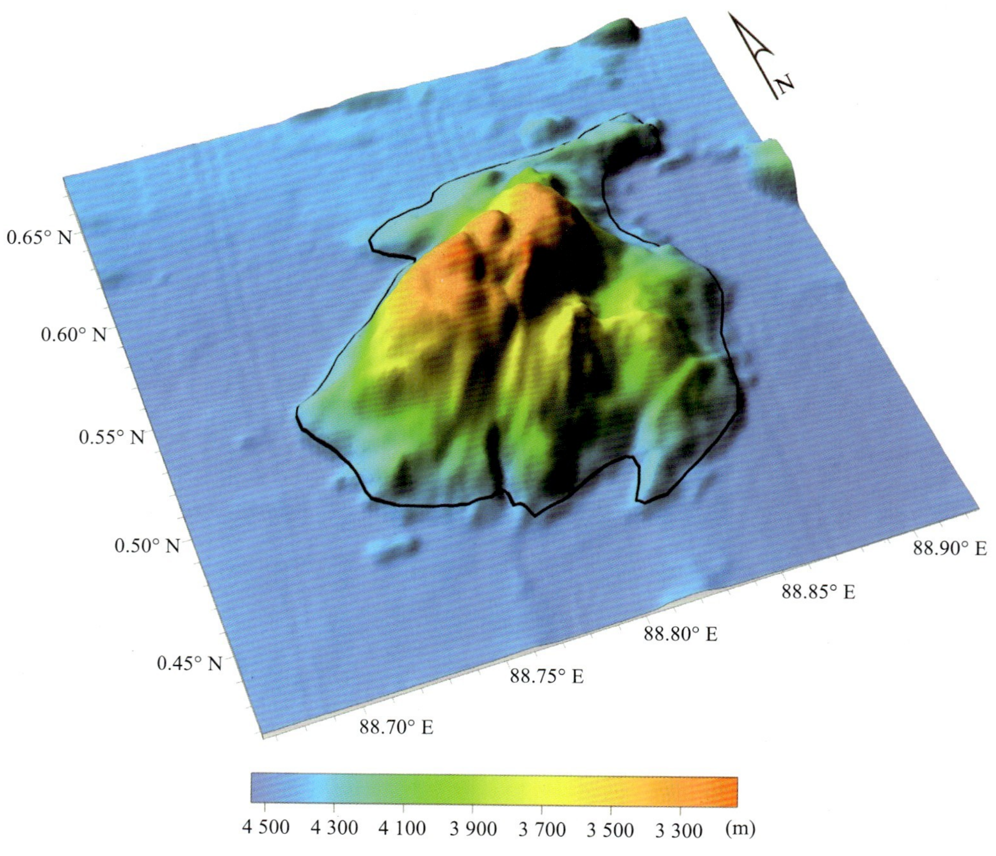

图 5-7　红袍海山：(a) 地形图（等深线间隔 200 m）；(b) 三维图

Fig.5-7　Hongpao Seamount: (a) Bathymetric map (Contours are in 200 m); (b) 3-D topographic map

5.2.7 猴魁海丘 Houkui Hill

中文名称 Chinese Name	Hóukuí Hǎiqiū 猴魁海丘		
英文名称 English Name	Houkui Hill		
地理区域 / Location	东印度洋 The East Indian Ocean		
特征点坐标 Coordinate	00°39.7′N 90°37.5′E	长度 / Length 宽度 / Width	44 km 13 km
水深 / Depth	3 663~4 497 m	高差 / Total Relief	834 m
发现情况 Discovery Facts	此海丘于 2019 年“向阳红 01”船在执行航次调查时发现。 The hill was discovered in 2019 during the survey cruise carried out onboard the Chinese R/V *Xiang Yang Hong 01*.		
地形特征 Feature Description	猴魁海丘发育于东印度洋东经九十度海岭以东 18km，呈 NE-SW 向条状延伸，东西长约 44 km，南北宽约 13 km。该海丘整体发育多个小规模凸起，最高处位于海丘西部，水深约 3 663 m。 The hill is 18km east of the Ninety East Ridge in the East Indian Ocean. Extending in the NE-SW direction like a stripe, it is about 44 km long from west to east, and about 13 km wide from south to north. There are many small protrusions on the hill, and the highest point is located in the west of the hill, with a depth of about 3,663 m.		
专名释义 Reason for Choice of Name	该区域采用中国茶文化群组化方法命名，以象征中国通过印度洋与西方各国友好的贸易和文化往来。此海丘以猴魁命名。猴魁茶是一种汉族传统名茶，产于中国安徽黄山市。 The features in this sea area are all named following the Chinese tea culture naming system to symbolize the friendly trade and cultural exchanges between China and Western countries through the Indian Ocean. The hill is named Houkui Hill. Houkui is the name of a well-known Chinese tea produced in Huangshan City, Anhui Province, China.		

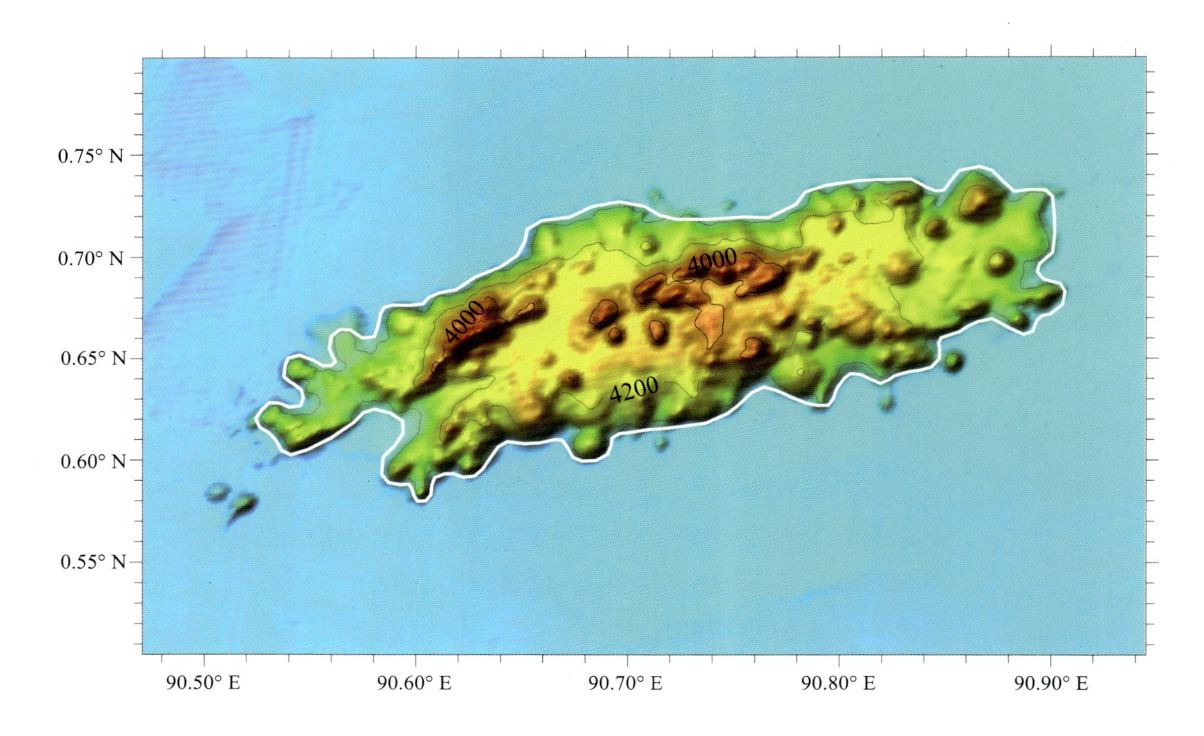

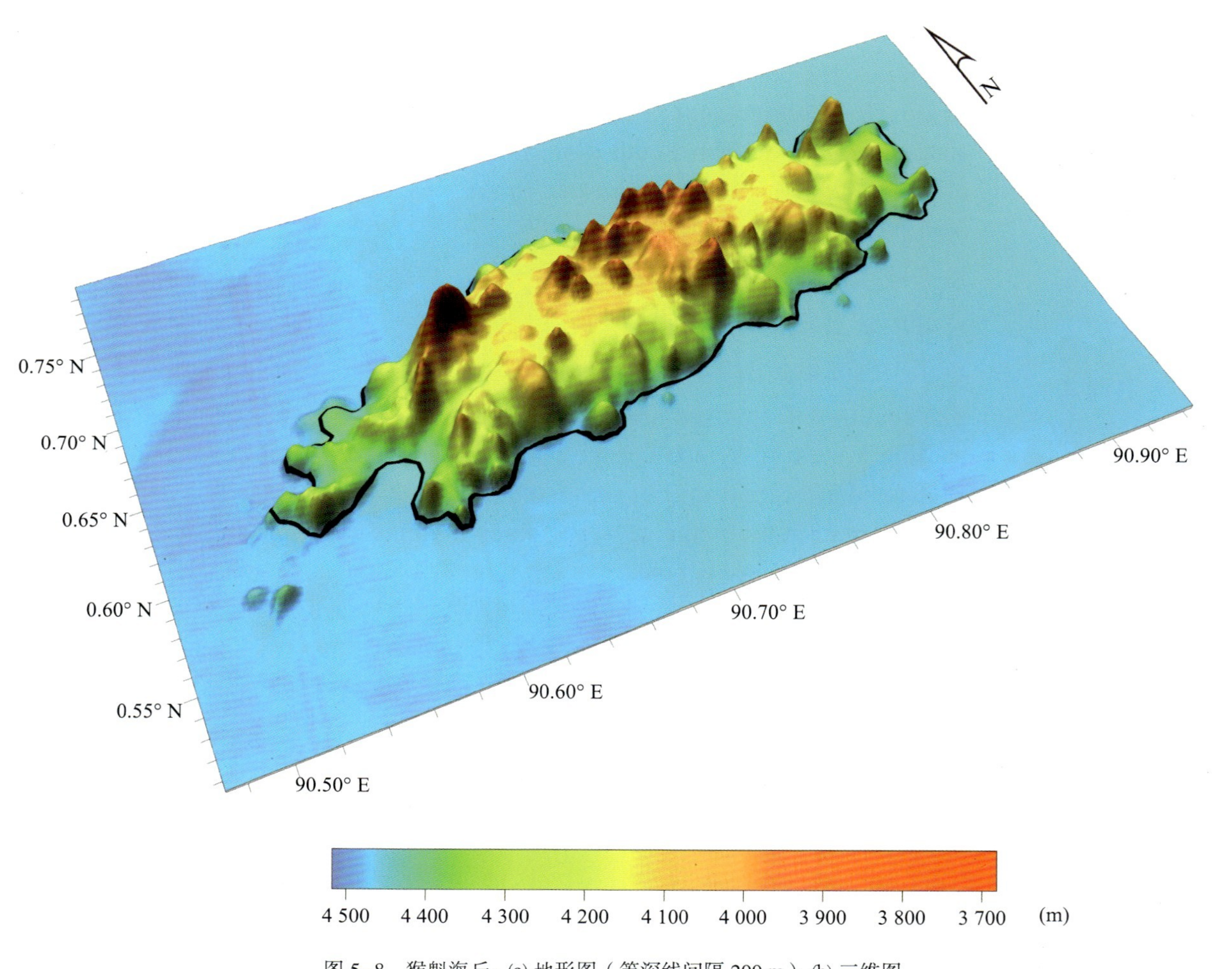

图 5-8 猴魁海丘：(a) 地形图（等深线间隔 200 m）；(b) 三维图

Fig.5-8 Houkui Hill: (a) Bathymetric map (Contours are in 200 m); (b) 3-D topographic ma

5.2.8 茗眉海脊 Mingmei Ridge

中文名称 Chinese Name	Míngméi Hǎijǐ 茗眉海脊		
英文名称 English Name	Mingmei Ridge		
地理区域 / Location	东印度洋 The East Indian Ocean		
特征点坐标 Coordinate	00°22.0′N 90°14.8′E	长度 / Length 宽度 / Width	76 km 19 km
水深 / Depth	2 679~4 596 m	高差 / Total Relief	1 917 m
发现情况 Discovery Facts	此海脊于 2019 年"向阳红 01"船在执行航次调查时发现。 The ridge was discovered in 2019 during the survey cruise carried out onboard the Chinese R/V *Xiang Yang Hong 01*.		
地形特征 Feature Description	茗眉海脊发育于东印度洋东经九十度海岭以东 12 km，整体呈 NE-SW 向带状延伸，长约 76 km，宽约 19 km。该海脊顶部高耸，两侧边缘陡峭，最高处位于海脊东北部，水深约 2 679 m。 It is 12 km east of the Ninety East Ridge in the East Indian Ocean. Extending in the NE-SW direction like a belt as a whole, it is about 76 km long and about 19 km wide. With a towering top, the ridge has steep slopes on the both sides. The highest point is in the northeast of the ridge, with a depth of about 2,679 m.		
专名释义 Reason for Choice of Name	该区域采用中国茶文化群组化方法命名，以象征中国通过印度洋与西方各国友好的贸易和文化往来。茗眉茶，属于绿茶，产于中国江西婺源一带。此海脊西南端圆润，如侍女眉首，东北两条山脊分叉如眉尾，故以茗眉命名。 The features in this sea area are named following the Chinese tea culture naming system to symbolize the friendly trade and cultural exchanges between China and Western countries through the Indian Ocean. The Mingmei tea is a kind of green tea produced in Wuyuan, Jiangxi Province, China. The southwestern end of the ridge and the two diverging ridges in the northeast look like a leaf of the tea. Hence, the feature is named Mingmei Ridge.		

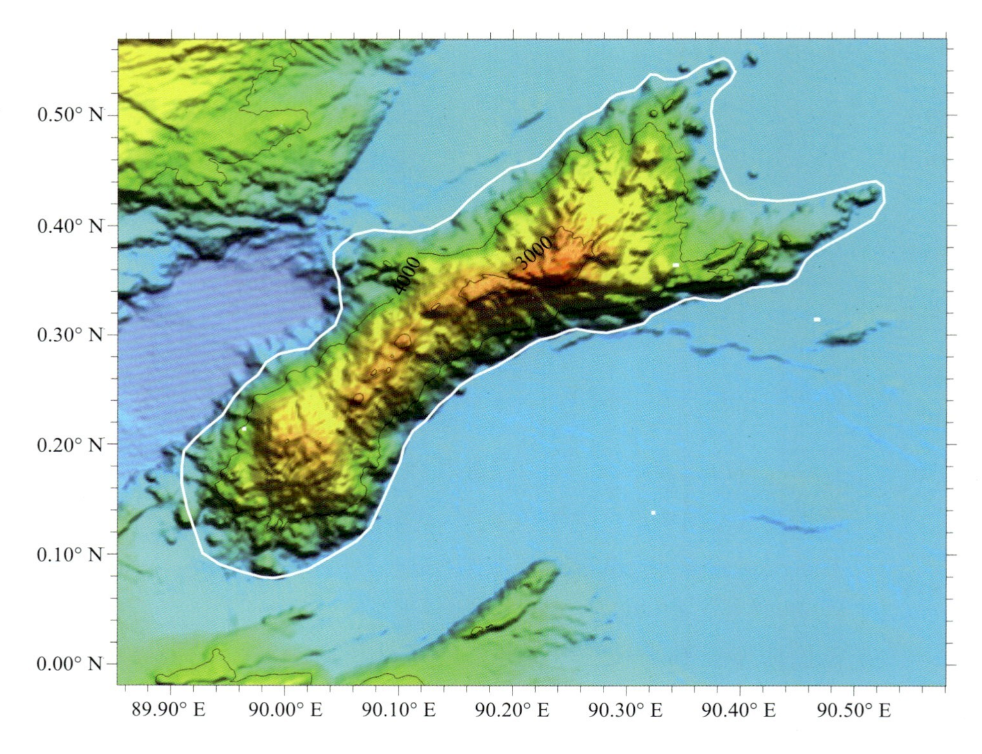

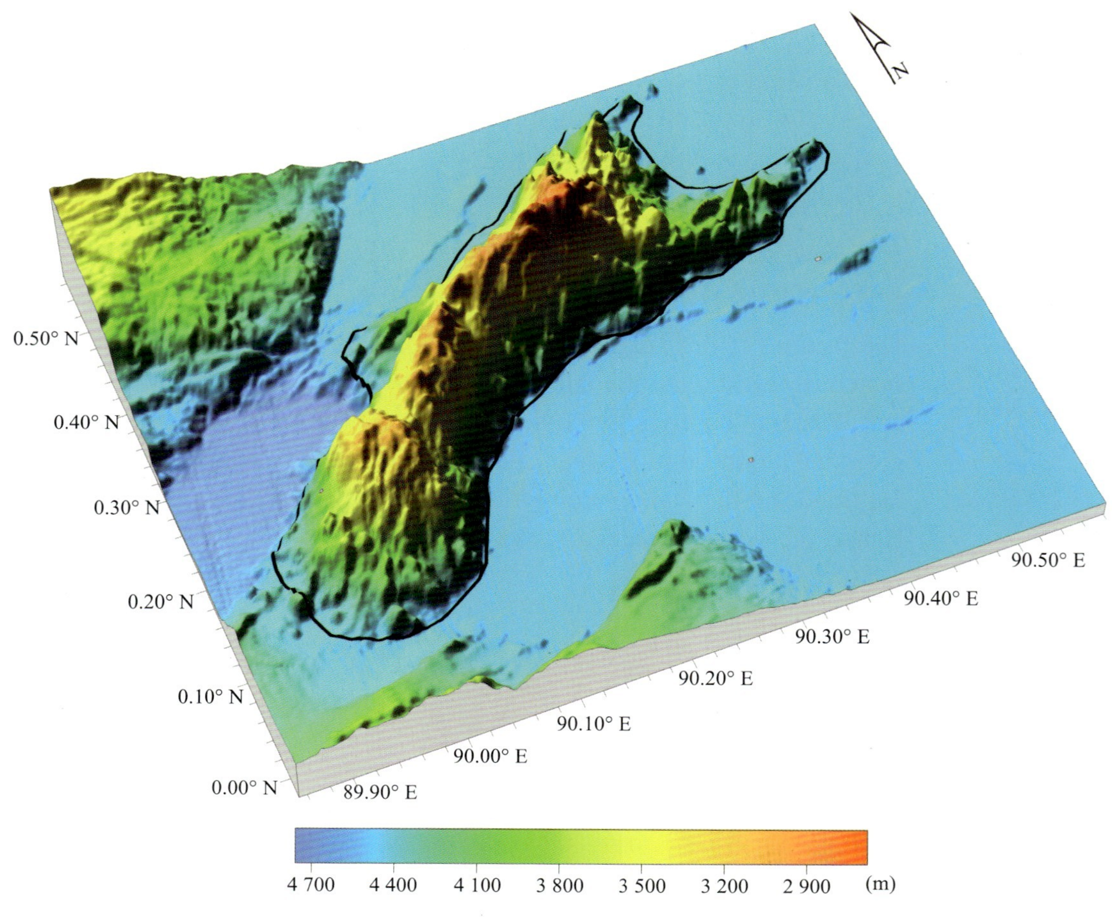

图 5-9 茗眉海脊：(a) 地形图（等深线间隔 1 000 m）；(b) 三维图

Fig.5-9 Mingmei Ridge: (a) Bathymetric map (Contours are in 1,000 m); (b) 3-D topographic map

5.2.9 普洱海山群 Pu’er Seamounts

<table>
<tr><td>中文名称
Chinese Name</td><td colspan="3">Pǔěr Hǎishānqún
普洱海山群</td></tr>
<tr><td>英文名称
English Name</td><td colspan="3">Pu’er Seamounts</td></tr>
<tr><td>地理区域 / Location</td><td colspan="3">东印度洋 The East Indian Ocean</td></tr>
<tr><td rowspan="2">特征点坐标
Coordinate</td><td rowspan="2">00°21.6′S 88°37.5′E
00°22.7′S 88°23.8′E</td><td>长度 / Length</td><td>41 km</td></tr>
<tr><td>宽度 / Width</td><td>15 km</td></tr>
<tr><td>水深 / Depth</td><td>2 510~4 449 m</td><td>高差 / Total Relief</td><td>1 939 m</td></tr>
<tr><td>发现情况
Discovery Facts</td><td colspan="3">此海山群于 2019 年“向阳红 01”船在执行航次调查时发现。
The seamounts were discovered in 2019 during the survey cruise carried out onboard the Chinese R/V Xiang Yang Hong 01.</td></tr>
<tr><td>地形特征
Feature Description</td><td colspan="3">普洱海山群发育于东印度洋，紧邻东经九十度海岭东侧。该海山群分布大小 2 个海山，最高处位于东部海山，峰顶水深 2 510 m，山麓水深 4 449 m。
Located in the East Indian Ocean, the seamounts are adjacent to the east of the Ninety East Ridge. There are 2 seamounts with difference sizes. The highest point is located in the eastern seamount, which is 2,510 m deep at the summit, and 4,449 m deep at the piedmont.</td></tr>
<tr><td>专名释义
Reason for Choice of Name</td><td colspan="3">该区域采用中国茶文化群组化方法命名，以象征中国通过印度洋与西方各国友好的贸易和文化往来。普洱，是我国云南的特产茶门类。故以普洱命名。
The features in this sea area are all named following the Chinese tea culture naming system to symbolize the friendly trade and cultural exchanges between China and Western countries through the Indian Ocean. Pu’er is a specialty tea produced in Yunnan Province, China. Hence, the features are named Pu’er Seamounts.</td></tr>
</table>

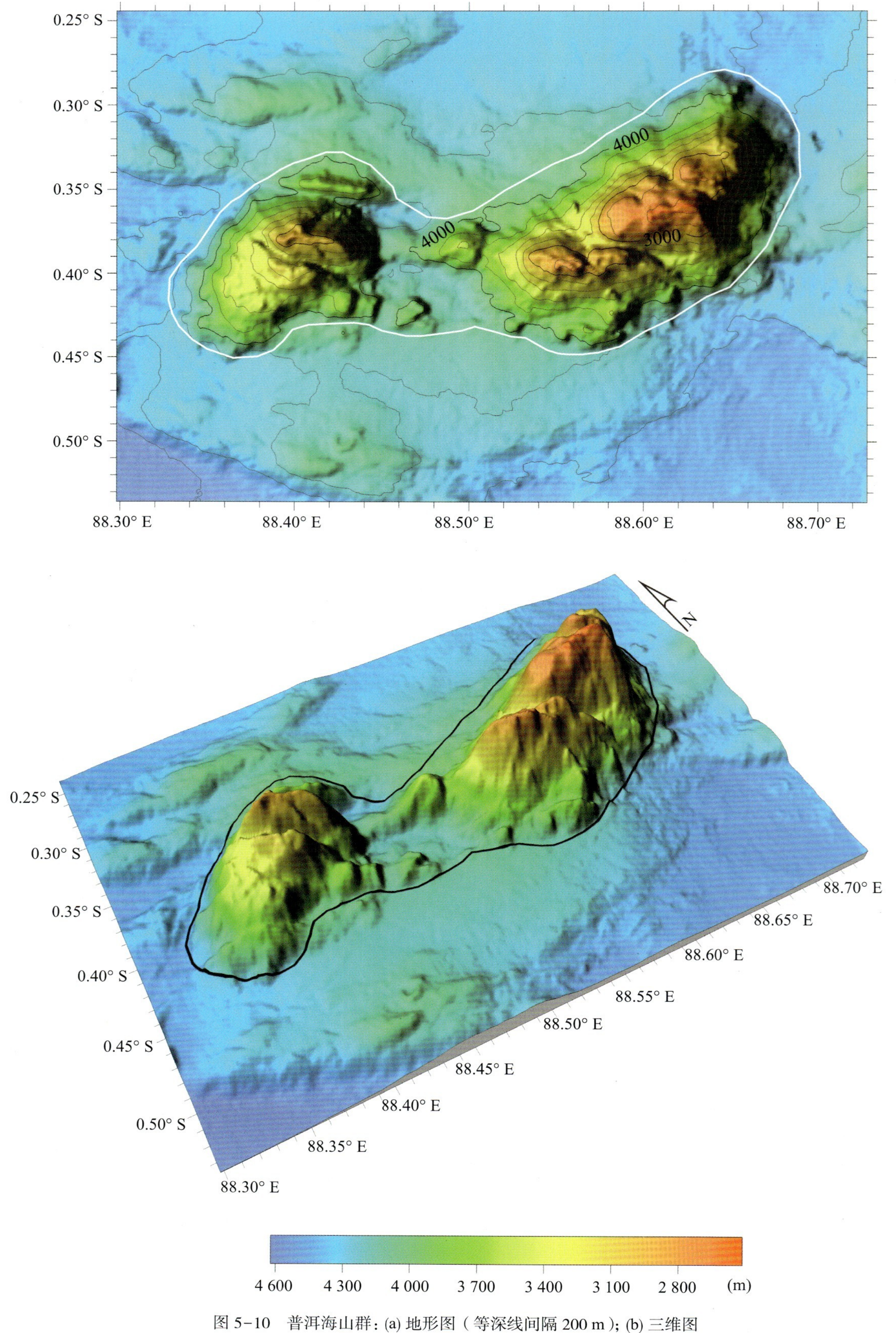

图 5-10　普洱海山群：(a) 地形图（等深线间隔 200 m）；(b) 三维图

Fig.5-10　Pu'er Seamounts: (a) Bathymetric map (Contours are in 200 m); (b) 3-D topographic map

5.2.10 雀舌海丘群 Queshe Hills

<table>
<tr><td>中文名称
Chinese Name</td><td colspan="3">Quèshé Hǎiqiūqún
雀舌海丘群</td></tr>
<tr><td>英文名称
English Name</td><td colspan="3">Queshe Hills</td></tr>
<tr><td>地理区域 / Location</td><td colspan="3">东印度洋 The East Indian Ocean</td></tr>
<tr><td rowspan="2">特征点坐标
Coordinate</td><td rowspan="2">00°57.6′S 88°18.9′E
00°57.6′S 88°10.0′E
00°58.4′S 88°17.4′E</td><td>长度 / Length</td><td>24 km</td></tr>
<tr><td>宽度 / Width</td><td>10 km</td></tr>
<tr><td>水深 / Depth</td><td>3 677~4 644 m</td><td>高差 / Total Relief</td><td>967 m</td></tr>
<tr><td>发现情况
Discovery Facts</td><td colspan="3">此海丘群于 2019 年“向阳红 01”船在执行航次调查时发现。
The hills were discovered in 2019 during the survey cruise carried out onboard the Chinese R/V Xiang Yang Hong 01.</td></tr>
<tr><td>地形特征
Feature Description</td><td colspan="3">雀舌海丘群位于东印度洋东经九十度海岭以西 35 km，呈 N-S 向延伸，东西长约 24 km，南北宽约 10 km。该海丘群发育 3 个规模相同的山体，最高处位于东部山体，水深约 3 677 m。
Located 35 km west of the Ninety East Ridge in the East Indian Ocean, the hills extend in the N-S direction, and are about 24 km long from west to east, and about 10 km wide from south to north. There are three hills of the same scale. The highest point is located in the middle, with a depth of about 3,677 m.</td></tr>
<tr><td>专名释义
Reason for Choice of Name</td><td colspan="3">该区域采用中国茶文化群组化方法命名，以象征中国通过印度洋与西方各国友好的贸易和文化往来。此海丘群以雀舌命名。雀舌，属绿茶，因形状小巧似雀舌而得名。
The features in this sea area are all named following the Chinese tea culture naming system to symbolize the friendly trade and cultural exchanges between China and Western countries through the Indian Ocean. The hills are named Queshe Hills. Queshe is the name of a kind of green tea with its leaf shaped like the tongue of a bird.</td></tr>
</table>

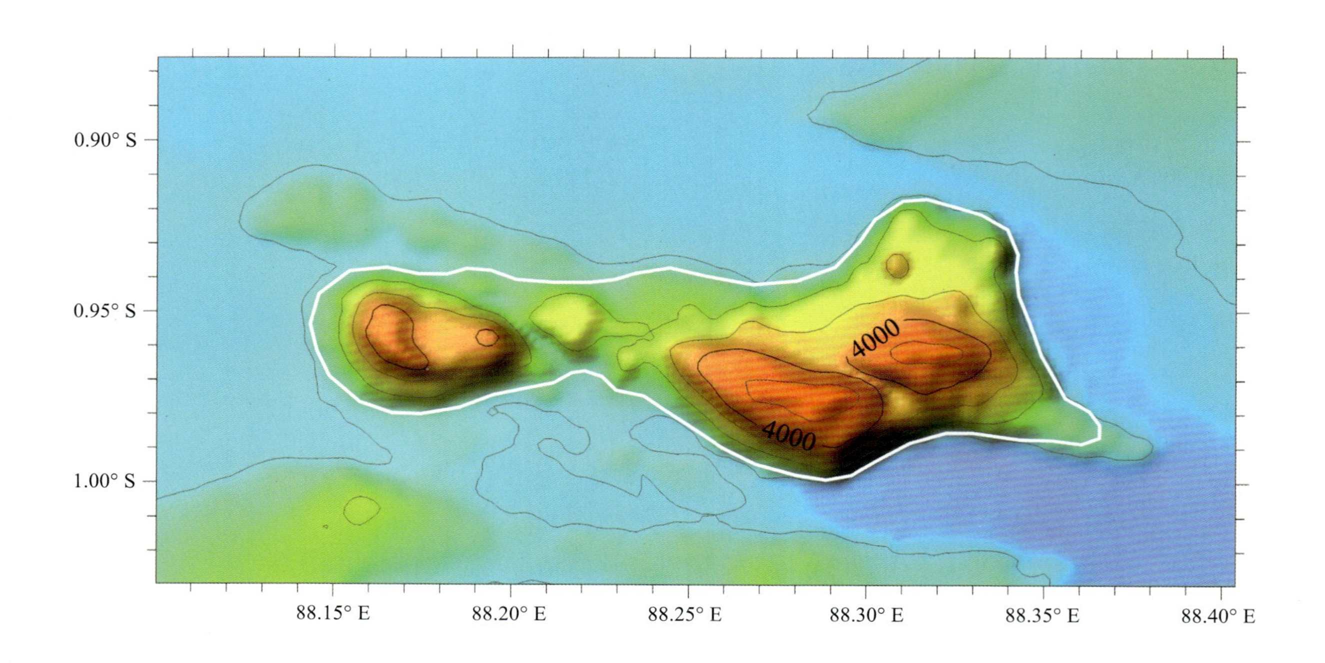

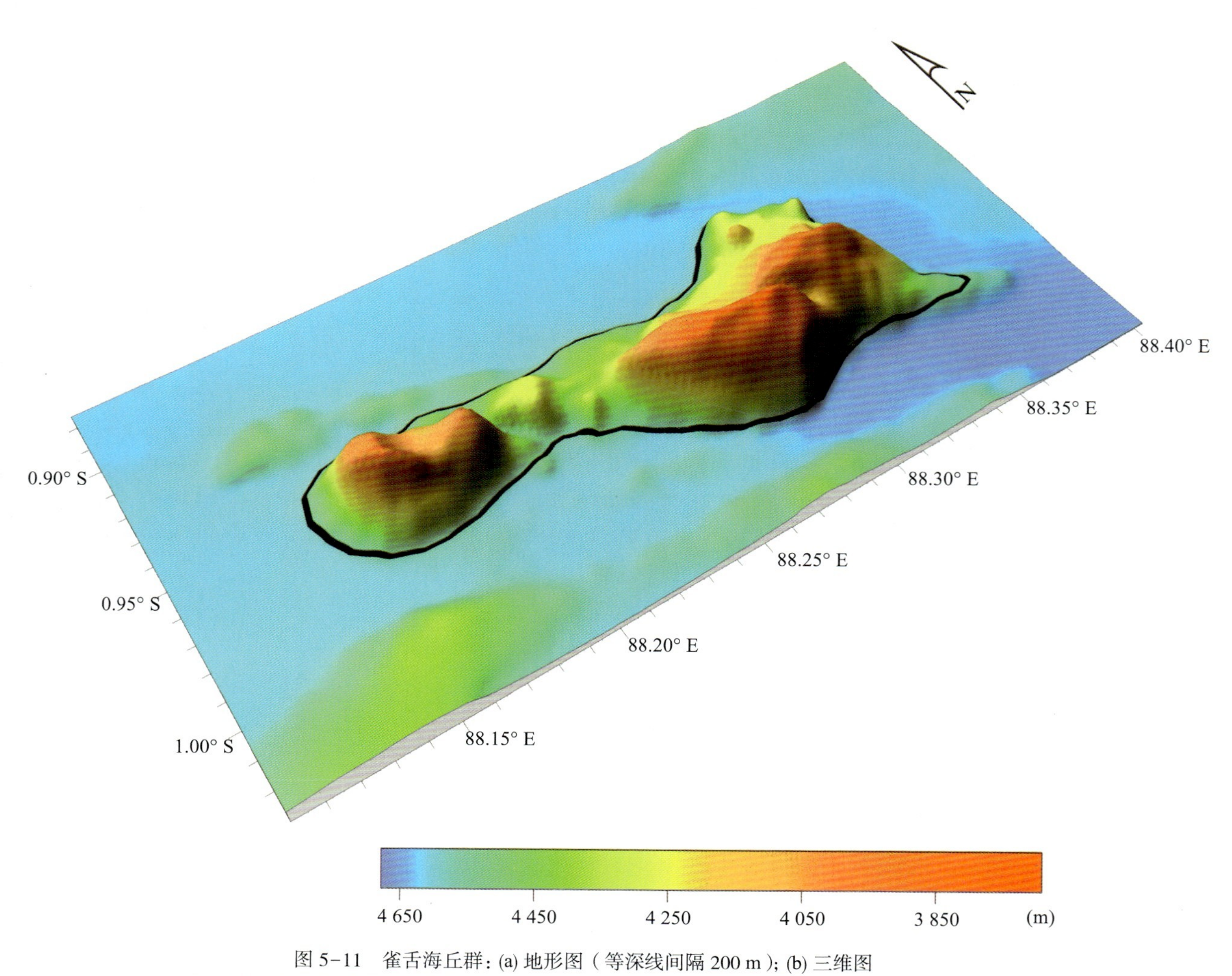

图 5-11　雀舌海丘群：(a) 地形图（等深线间隔 200 m）；(b) 三维图

Fig.5-11　Queshe Hills: (a) Bathymetric map (Contours are in 200 m); (b) 3-D topographic map

5.2.11 仙毫海脊 Xianhao Ridge

中文名称 Chinese Name	Xiānháo Hǎijǐ 仙毫海脊		
英文名称 English Name	Xianhao Ridge		
地理区域 / Location	东印度洋 The East Indian Ocean		
特征点坐标 Coordinate	01°58.1′S 89°46.1′E	长度 / Length	64 km
		宽度 / Width	17 km
水深 / Depth	2 993~4 709 m	高差 / Total Relief	1 716 m
发现情况 Discovery Facts	此海脊于 2019 年“向阳红 01”船在执行航次调查时发现。 The ridge was discovered in 2019 during the survey cruise carried out onboard the Chinese R/V *Xiang Yang Hong 01*.		
地形特征 Feature Description	仙毫海脊发育于东印度洋东经九十度海岭以西 13 km，整体呈 NE-SW 向带状延伸，长约 64 km，宽约 17 km。脊顶高耸，两侧陡峭，最高处位于海脊中部，水深 2 993 m。 Located 13 km west of the Ninety East Ridge in the East Indian Ocean, the ridge extends in the NE-SW direction as a whole, and it is about 64 km long and about 17 km wide. It is towering at the top and steep on the two sides. The highest point is located in the middle of the ridge, with a depth of 2,993 m.		
专名释义 Reason for Choice of Name	该区域采用中国茶文化群组化方法命名，以象征中国通过印度洋与西方各国友好的贸易和文化往来。仙毫茶，外形修长，产自中国陕西省汉中盆地。此海脊纤细修长，形似仙毫，故以仙毫命名。 The features in this area are all named following the Chinese tea culture naming system to symbolize the friendly trade and cultural exchanges between China and Western countries through the Indian Ocean. Xianhao is the name of a kind of tea produced in the Hanzhong Basin of Shanxi Province, China. This ridge is slender and shaped like the Xianhao tea leaf. Hence, the feature is named Xianhao Ridge.		

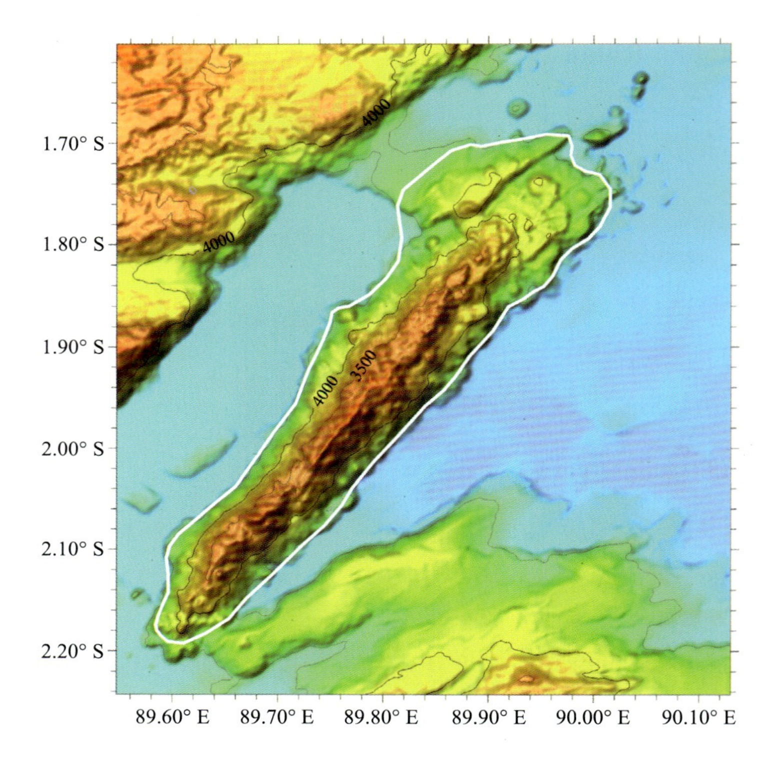

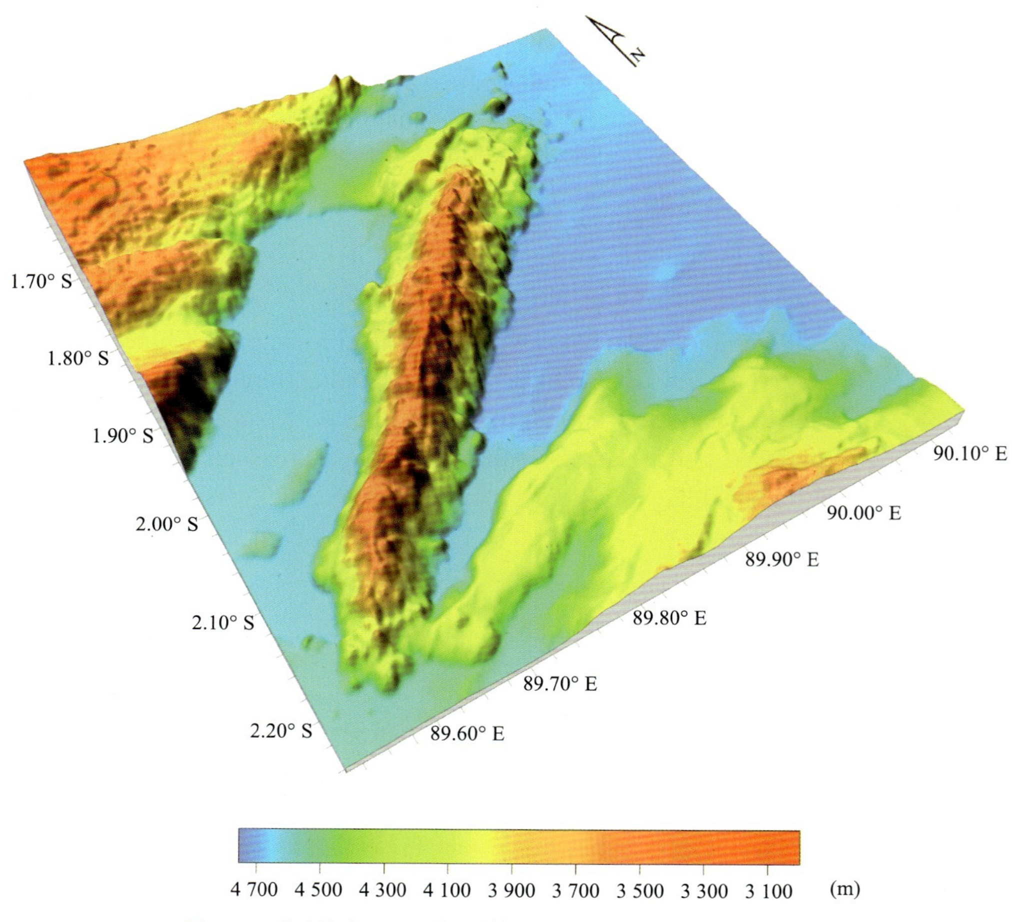

图 5-12 仙毫海脊：(a) 地形图（等深线间隔 500 m）；(b) 三维图
Fig.5-12 Xianhao Ridge: (a) Bathymetric map (Contours are in 500 m); (b) 3-D topographic map

5.2.12 岭西南海底水道 Lingxinan Sea Channel

中文名称 Chinese Name	Lǐngxīnán Hǎidǐshuǐdào 岭西南海底水道		
英文名称 English Name	Lingxinan Sea Channel		
地理区域 / Location	东印度洋 The East Indian Ocean		
特征点坐标 Coordinate	03°20.6′S 87°33.4′E	长度 / Length	142 km
		宽度 / Width	4 km
水深 / Depth	4 816~4 972 m	高差 / Total Relief	156 m
发现情况 Discovery Facts	此海底水道于 2019 年"向阳红 01"船在执行航次调查时发现。 The sea channel was discovered in 2019 during the survey cruise carried out onboard the Chinese R/V *Xiang Yang Hong 01*.		
地形特征 Feature Description	岭西南海底水道发育于东印度洋东经九十度海岭以西 60 km，呈 NE-SW 向弯曲延伸，南北长超过 140 km，东西宽约 4 km，平均高差约 156 m。 Located 60 km west of the Ninety East Ridge in the East Indian Ocean, the sea channel extends in the NE-SW direction, and is more than 140 km long from south to north and about 4 km wide from west to east. The average total relief is about 156 m.		
专名释义 Reason for Choice of Name	该海底水道位于东经九十度海岭西南，故以岭西南命名。 The sea channel is located southwest of the Ninety East Ridge in the East Indian Ocean. The southwest of the ridge means Lingxinan in Chinese, hence the feature is named Lingxinan Sea Channel.		

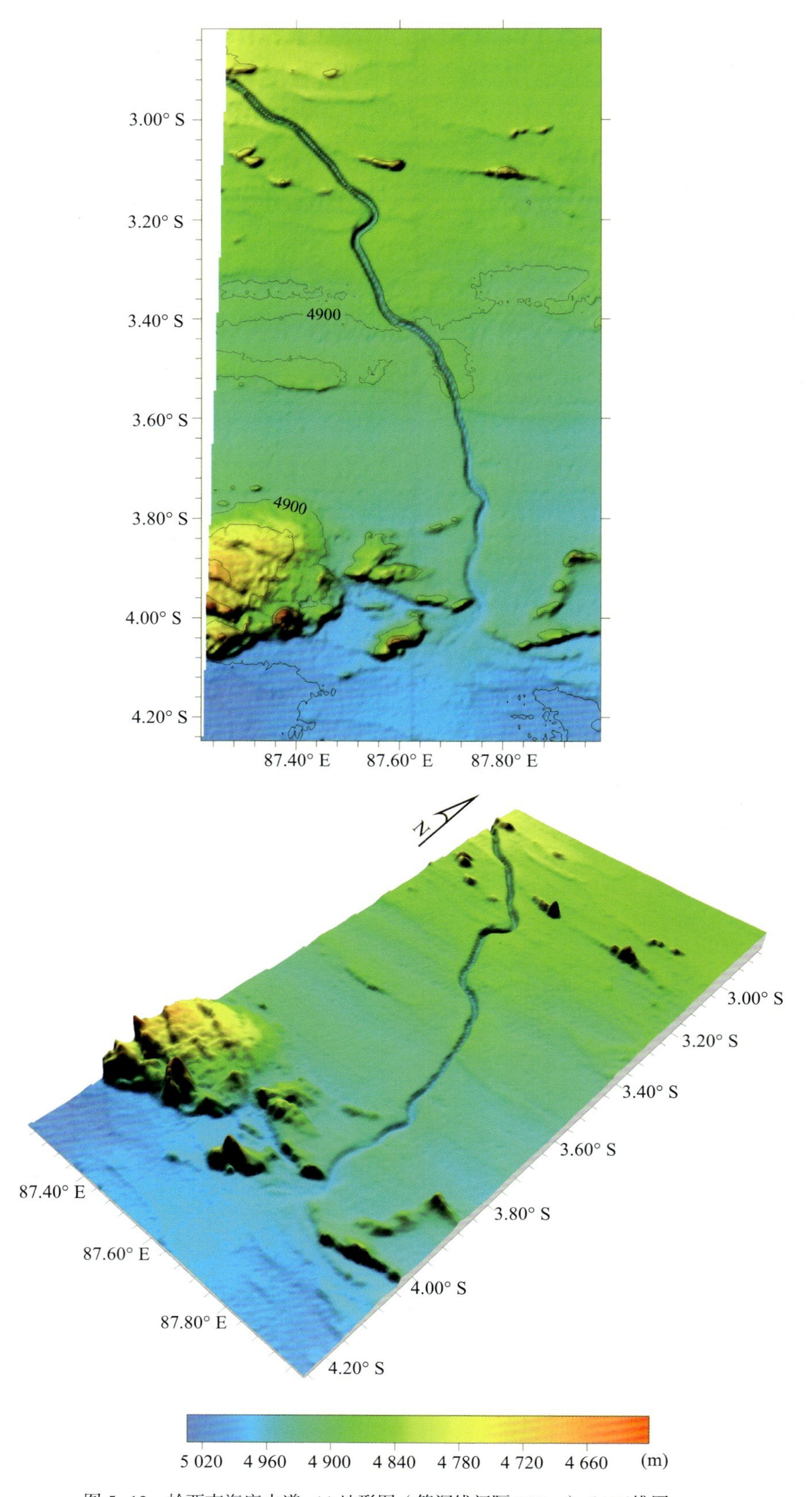

图 5-13　岭西南海底水道：(a) 地形图（等深线间隔 100 m）；(b) 三维图

Fig.5-13　Lingxinan Sea Channel: (a) Bathymetric map (Contours are in 100 m); (b) 3-D topographic map

5.2.13 泉茗海丘 Quanming Hill

中文名称 Chinese Name	Quánmíng Hǎiqiū 泉茗海丘		
英文名称 English Name	Quanming Hill		
地理区域 / Location	东印度洋 The East Indian Ocean		
特征点坐标 Coordinate	03°55.7′S 88°16.1′E	长度 / Length	9.3 km
		宽度 / Width	8.7 km
水深 / Depth	4 255~4 941 m	高差 / Total Relief	686 m
发现情况 Discovery Facts	此海丘于 2019 年“向阳红 01”船在执行航次调查时发现。 The hill was discovered in 2019 during the survey cruise carried out onboard the Chinese R/V *Xiang Yang Hong 01.*		
地形特征 Feature Description	泉茗海丘发育于东印度洋东经九十度海岭以西 12 km，整体长约 9.3 km，宽约 8.7 km。该海丘山顶呈马鞍形，最高处位于山顶西南部，水深约 4 255 m，东南山麓平缓，西北陡峭。 Located 12 km west of the Ninety East Ridge in the East Indian Ocean, the hill is about 9.3 km long and about 8.7 km wide as a whole. The hill has a saddle-shaped summit, and the highest point is located in the southwest of the summit, with a depth of about 4,255 m. The foot of the hill in the southeast is gentle, whereas that of the hill in the northwest is steep.		
专名释义 Reason for Choice of Name	该区域采用中国茶文化群组化方法命名，以象征中国通过印度洋与西方各国友好的贸易和文化往来。泉茗，属绿茶，是陕西商洛的一种特产茶。故以商洛命名。 The features in the area are all named following the Chinese tea culture naming system to symbolize the friendly trade and cultural exchanges between China and Western countries through the Indian Ocean. Quanming is the name of a kind of specialty green tea produced in Shangluo, Shanxi Province, China. Hence, the feature is named Quanming Hill.		

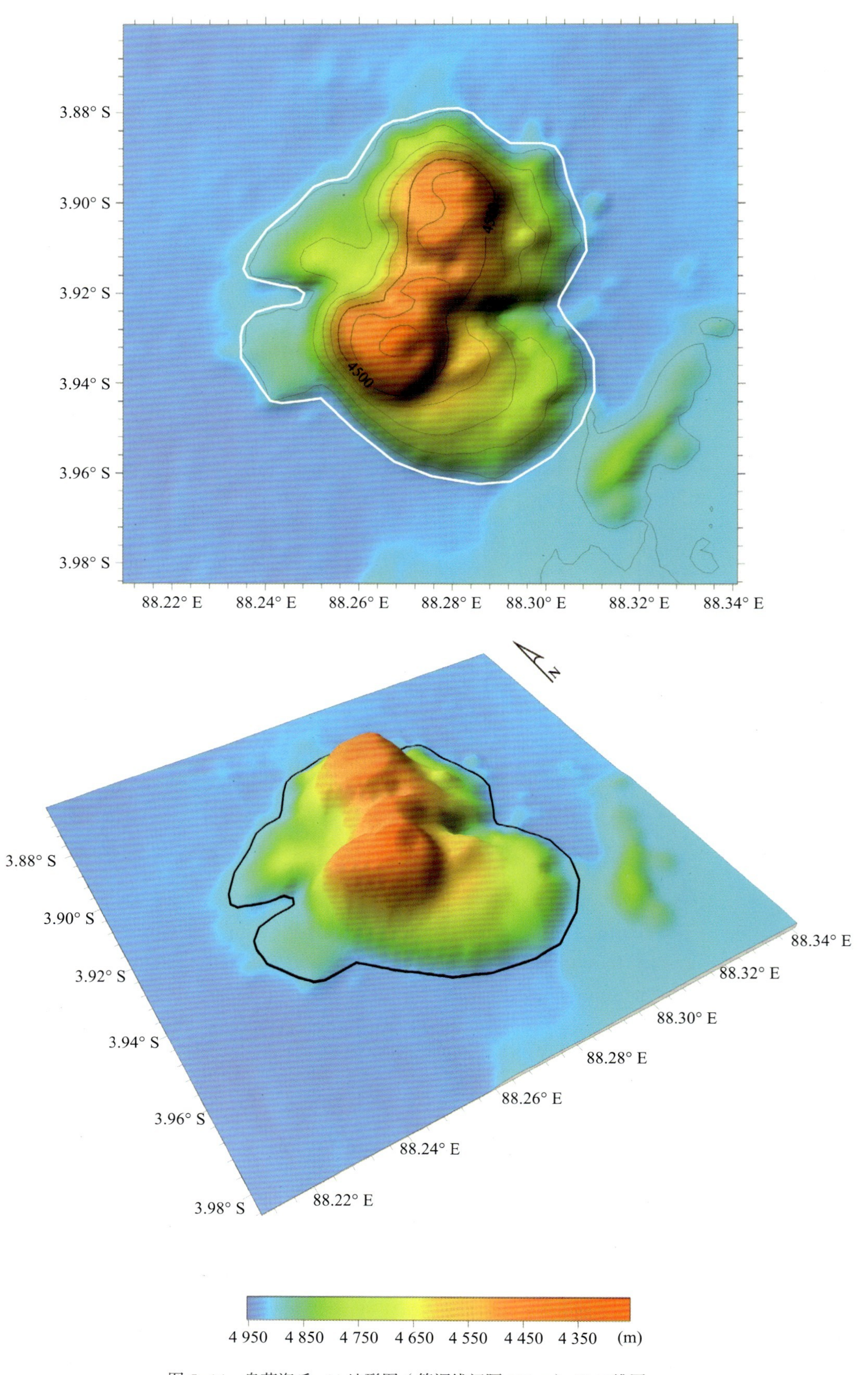

图 5-14　泉茗海丘：(a) 地形图（等深线间隔 100 m）；(b) 三维图

Fig.5-14　Quanming Hill: (a) Bathymetric map (Contours are in 100 m); (b) 3-D topographic map

5.2.14 银针海脊 Yinzhen Ridge

中文名称 Chinese Name	Yínzhēn Hǎijǐ 银针海脊		
英文名称 English Name	Yinzhen Ridge		
地理区域 / Location	东印度洋 The East Indian Ocean		
特征点坐标 Coordinate	03°35.5′S 90°45.5′E	长度 / Length 宽度 / Width	140 km 7.8 km
水深 / Depth	2 948~4 974 m	高差 / Total Relief	2 026 m
发现情况 Discovery Facts	此海脊于 2019 年“向阳红 01”船在执行航次调查时发现。 The ridge was discovered in 2019 during the survey cruise carried out onboard the Chinese R/V *Xiang Yang Hong 01.*		
地形特征 Feature Description	银针海脊发育于东印度洋东经九十度海岭以东 7 km，呈 N-S 向直线延伸，南北长约 140 km，东西宽约 7.8 km。该海脊两头尖锐，北高南低，最高处位于脊顶中北部，水深 2 948 m，中部略宽，侧翼陡峭。 Located 7 km east of the Ninety East Ridge in the East Indian Ocean, the ridge extends in the N-S direction like a straight line, and is about 140 km long from south to north and about 7.8 km wide from west to east. The two ends of the ridge are sharp, and it is high in the north and low in the south. The highest point is located in the central north of the ridge summit, with a depth of 2,948 m. The middle part is slightly wider, and the flanks are steep.		
专名释义 Reason for Choice of Name	该区域采用中国茶文化群组化方法命名，以象征中国通过印度洋与西方各国友好的贸易和文化往来。银针，属白茶，是我国福建的一种特产茶。该海脊细长如针，故以银针命名。 The features in this sea area are all named following the Chinese tea culture naming system to symbolize the friendly trade and cultural exchanges between China and Western countries through the Indian Ocean. Yinzhen (silver needle in Chinese) is the name of a kind of white tea produced in Fujian Province, China. The ridge is as long and thin shaped like a needle. Hence, the ridge is named Yinzhen Ridge.		

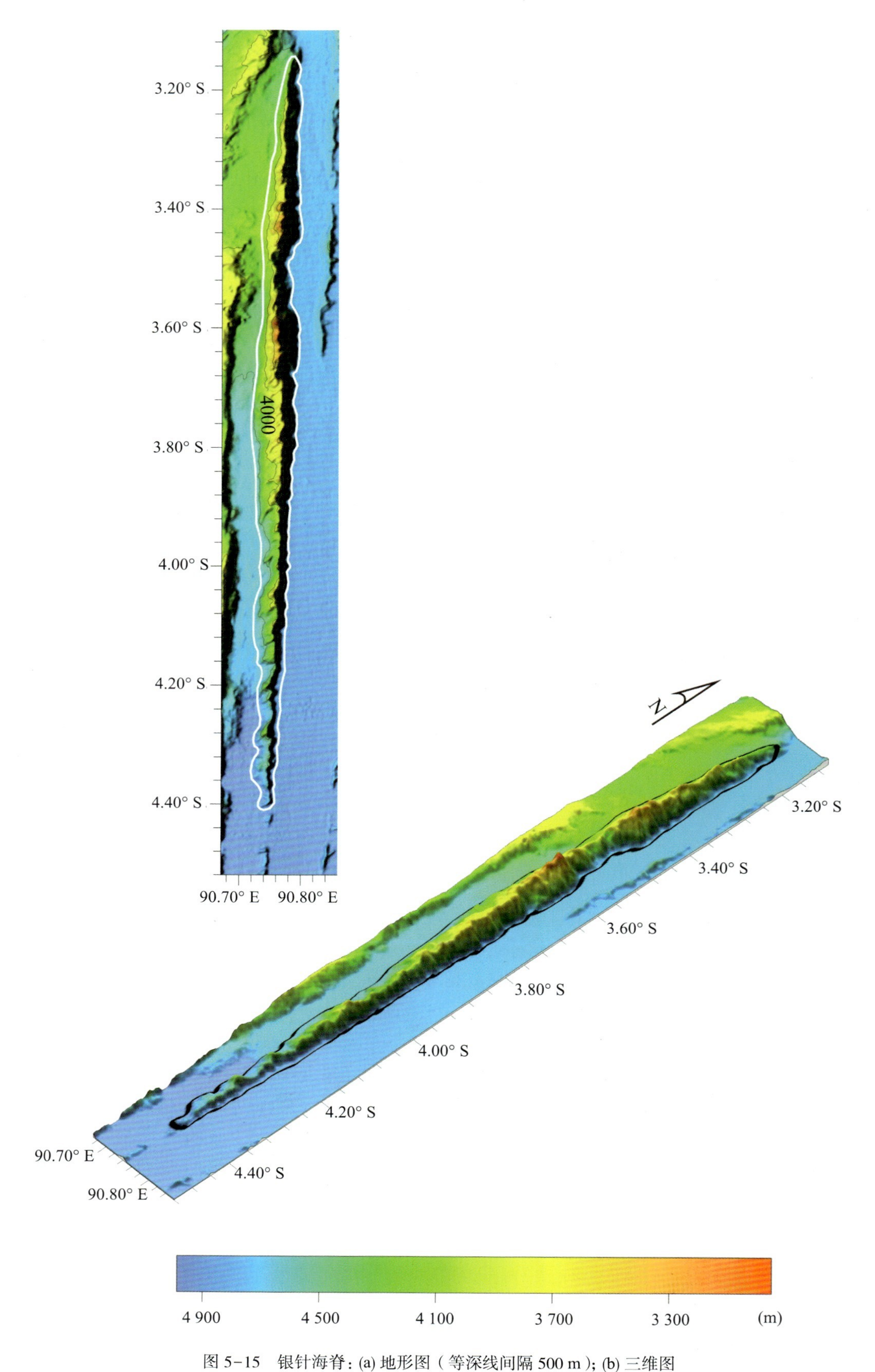

图 5-15　银针海脊：(a) 地形图（等深线间隔 500 m）；(b) 三维图

Fig.5-15　Yinzhen Ridge: (a) Bathymetric map (Contours are in 500 m); (b) 3-D topographic map

5.2.15 银梭海丘 Yinsuo Hill

<table>
<tr><td>中文名称
Chinese Name</td><td colspan="3">Yínsuō Hǎiqiū
银梭海丘</td></tr>
<tr><td>英文名称
English Name</td><td colspan="3">Yinsuo Hill</td></tr>
<tr><td>地理区域 / Location</td><td colspan="3">东印度洋 The East Indian Ocean</td></tr>
<tr><td rowspan="2">特征点坐标
Coordinate</td><td rowspan="2">04°56.1′S 90°28.2′E</td><td>长度 / Length</td><td>18 km</td></tr>
<tr><td>宽度 / Width</td><td>9 km</td></tr>
<tr><td>水深 / Depth</td><td>4 087~5 028 m</td><td>高差 / Total Relief</td><td>941 m</td></tr>
<tr><td>发现情况
Discovery Facts</td><td colspan="3">此海丘于 2019 年“向阳红 01”船在执行航次调查时发现。
The hill was discovered in 2019 during the survey cruise carried out onboard the Chinese R/V Xiang Yang Hong 01.</td></tr>
<tr><td>地形特征
Feature Description</td><td colspan="3">银梭海丘发育于东印度洋东经九十度海岭以东约 20 km，总长约 18 km，宽约 9 km。水深范围 4 087~5 028 m，最大高差约 941 m，位于海丘西部。
The hill is about 20 km east of the Ninety East Ridge in the East Indian Ocean, with a total length of about 18 km and a width of about 9 km. Its depths vary from 4,087 m to 5,082 m, with a maximum total relief of about 941 m located in the west of the hill.</td></tr>
<tr><td>专名释义
Reason for
Choice of Name</td><td colspan="3">该区域采用中国茶文化群组化方法命名，以象征中国通过印度洋与西方各国友好的贸易和文化往来。该海丘两端不一，形似我国梭状的银梭茶叶，故以银梭命名。
The features in this area are all named following the Chinese tea culture naming system to symbolize the friendly trade and cultural exchanges between China and Western countries through the Indian Ocean. The hill is shape as a leaf of the tea called Yinsuo. Hence, the hill is named Yinsuo Hill.</td></tr>
</table>

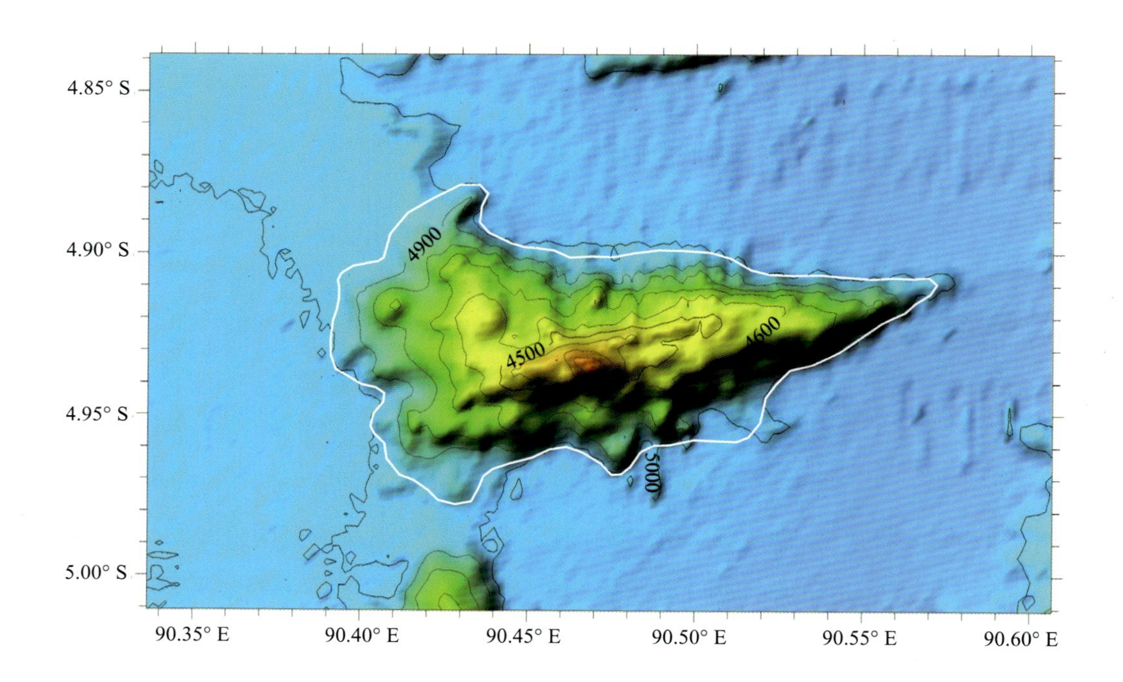

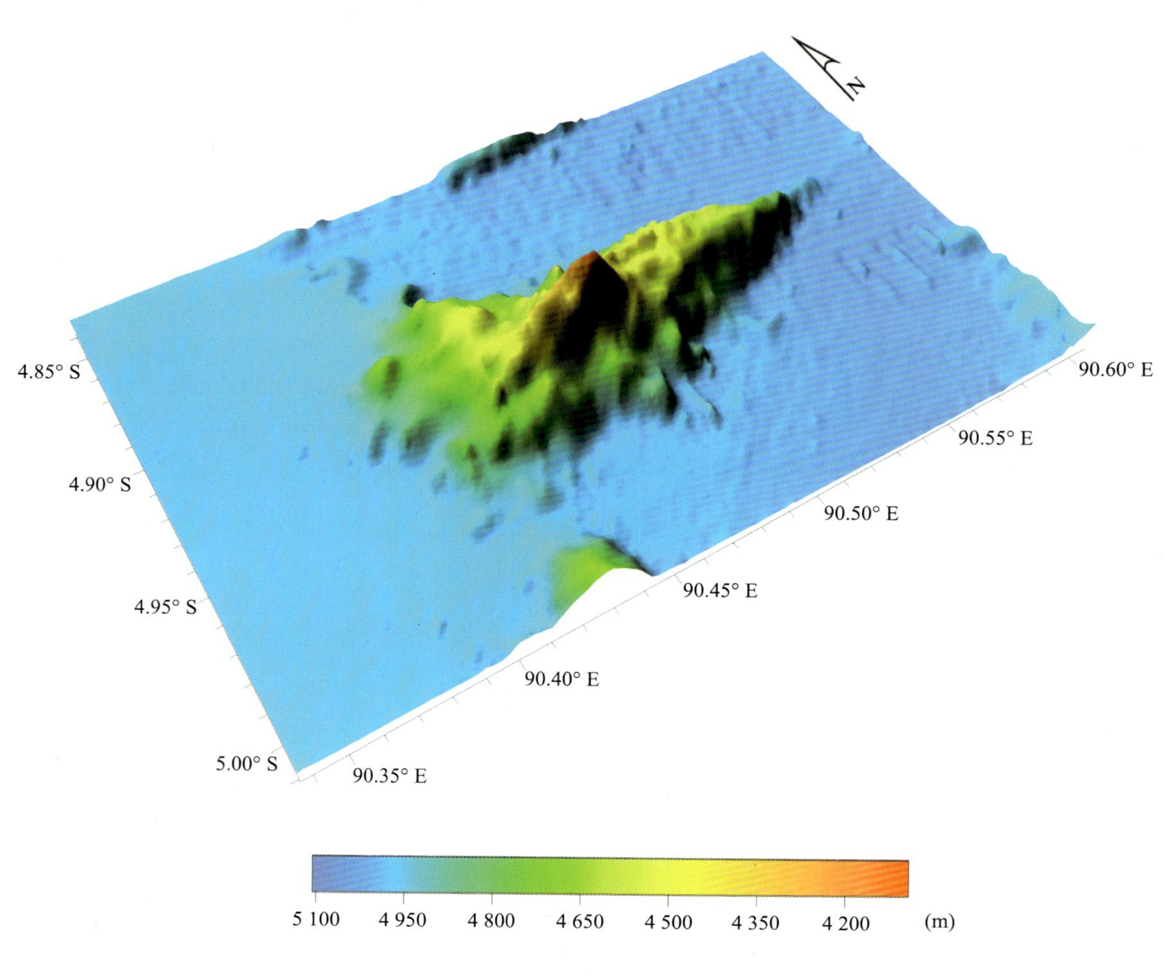

图 5-16　银梭海丘：(a) 地形图（等深线间隔 100 m）；(b) 三维图

Fig.5-16　Yinsuo Hill: (a) Bathymetric map (Contours are in 100 m); (b) 3-D topographic map

参考文献
References

[1] 张有学，尹安 . 地球的结构、演化和动力学 [M]. 北京：高等教育出版社，2002.

[2] 琼斯 . 海洋地球物理 [M]. 金翔龙，等，译 . 北京：海洋出版社，2009.

[3] 汪榕培，任秀桦 . 诗经（中英文版）[M]. 沈阳：辽宁教育出版社，1995.

[4] 李四海 , 邢喆 , 李艳雯 , 等 . 海底地理实体命名研究进展与发展趋势 [J]. 海洋通报，2012,31(5):594–600.

[5] 国际水道测量组织，政府间海洋学委员会 . 海底地名命名标准：B–6 出版物 4.1.0 版 . 国家海洋信息中心，译 . 北京：海洋出版社，2014.

[6] 沙忠利，等 . 西太平洋深海化能生态系统大型生物图谱 [M]. 北京：科学出版社，2019.

[7] 李四海，邢喆，樊妙，等 . 海底地名命名理论与技术方法 [M]. 北京：海洋出版社，2015.

[8] 徐奎栋，等 . 西太平洋海沟洋脊交联区海山动物原色图谱 [M]. 北京：科学出版社，2020.

[9] 齐锐，万昊宜 . 漫步中国星空 [M]. 北京：科学普及出版社，2014.

[10] 刘蓝 . 中国古代十大音乐家 [M]. 昆明：云南大学出版社，2017.

[11] 陈椽 . 茶业通史 [M]. 北京：农业出版社，1984.

[12] Winterer E L , Metzler C V. Origin and subsidence of guyots in Mid-Pacific Mountains[J]. *Journal of Geophysical Research*, 1984, *89*(B12):9969-9979.

[13] Hall-Spencer J, Rogers A, Davies J, et al. Deep-Sea coral distribution on seamounts,oceanic islands,and continental slopes in the Northeast Atlantic[J]. *Bulletin of Marine Science*, 2007, *81*(Suppl.1):135-146.

[14] Hubbs CL. Initial discoveries of fish faunas on seamounts and offshore banks in the Eastern Pacific[J]. *Pacific Science*, 1959, *13*(4):311-316.

[15] Lavelle JW, Baker ET. Ocean currents at axial volcano,a Northeastern Pacific Seamount[J]. *Journal of Geophysical Research*, 2003, *108*(C2):3020.

[16] Yesson C, Clark MR, Taylor ML, et al. The global distribution of seamounts based on 30 arc seconds bathymetry data[J]. *Deep-Sea Research*, 2011, *58*(4):442-453.

[17] Kaneda K, Kodair S, Nishizawa A, et al. Structural evolution of preexisting oceanic crust through intraplate igneous activities in the Marcus-Wake seamount chain[J]. *Geochemistry Geophysics Geosystems*, 2010, *11*(10):69-71.

索引
Index

图书在版编目（C I P）数据

深海典型海域海底地理实体命名 : 英文、汉文 / 石绥祥, 樊妙, 邢喆主编. -- 青岛 : 中国海洋大学出版社, 2024. 12. -- ISBN 978-7-5670-3946-9

Ⅰ. P737.2-62

中国国家版本馆CIP数据核字第2024WH5919号

书　　名　深海典型海域海底地理实体命名
出版发行　中国海洋大学出版社
社　　址　青岛市香港东路23号　　　邮政编码　266071
出 版 人　刘文菁
网　　址　http: //pub.ouc.edu.cn
订购电话　0532-82032573（传真）
责任编辑　杨亦飞　郝倩倩　　　电　　话　0532-85902342
照　　排　青岛光合时代传媒有限公司
印　　制　青岛海蓝印刷有限责任公司
版　　次　2024年12月第1版
印　　次　2024年12月第1次印刷
成品尺寸　210 mm × 285 mm
印　　张　13.5
印　　数　1 ~ 1 000
字　　数　233千
定　　价　368.00元
审 图 号　GS鲁（2024）0485号